GEOLOGY AND PETROLOGY
OF THE GALÁPAGOS ISLANDS

The Geological Society of America, Inc.
Memoir 118

Geology and Petrology
of the Galápagos Islands

A. R. MCBIRNEY
University of Oregon
Eugene, Oregon

HOWEL WILLIAMS
University of California
Berkeley, California

With chemical analyses by Ken-ichiro Aoki

1969

Published by
THE GEOLOGICAL SOCIETY OF AMERICA, INC.
Colorado Building, P.O. Box 1719
Boulder, Colorado 80302

Printed in The United States of America

*The printing of this volume has been made possible
through the bequest of
Richard Alexander Fullerton Penrose, Jr.,
and is partially supported by a grant from
The National Science Foundation.*

Introduction and Acknowledgments

We were extremely fortunate to have taken part in the Galápagos International Scientific Project conducted by the University of California Extension Division with the support of National Science Foundation Grant GE-2370 and in cooperation with the Charles Darwin Foundation. We are indebted first to the late Professor R. L. Usinger, Director, and to Dr. R. I. Bowman, Co-Director of the expedition, for inviting us to be members, and to Dr. David Snow, then Director of the Darwin Station, for support and facilities in the course of our field work.

More than sixty scientists took part in the expedition, almost all, of course, being biologists. Among those who accompanied us were Professor J. C. Granja of the Department of Geology of the University of Ecuador, Quito; our paleontological colleagues, Professor J. Wyatt Durham and Dr. Victor Zullo; and Dr. Allan Cox of the U. S. Geological Survey, a geophysicist specializing in paleomagnetic studies, who worked closely with us on all but two islands of the archipelago and supplied us with information concerning Wenman and Culpepper islands, which we were unable to visit. To each of these gentlemen we express sincere thanks; their valuable contributions will become apparent on later pages.

Our ship, the *Golden Bear,* while on a training cruise of the California Maritime Academy, anchored in Academy Bay on Indefatigable Island on January 19, 1964; she returned from Tahiti to take us to Guayaquil, Ecuador, on February 28. We recall with pleasure the many courtesies of her officers and crew. During the first two weeks of our stay in the archipelago, we had the invaluable support of the U.S.S. *Pine Island* and her helicopters. Thanks to them, Dr. Allan Cox was able to land on the otherwise almost inaccessible islands of Wenman and Culpepper, and all three of us flew to Cape Berkeley and Tagus Cove on Albemarle Island and to Narborough, Abingdon, Bindloe, James, Jervis, and Duncan islands. What we did in a few days would have taken many weeks or even months without their help. Their support was of inestimable value. We also wish to

vii

thank the Ecuadorian Navy for the use of one of their patrol boats that took us to Hood, Charles, and Barrington islands and David Balfour for sailing us to Chatham Island.

The kindly assistance given by Dr. David Snow, then Director of the Darwin Station at Academy Bay, is remembered with pleasure; so is the warm hospitality of the Wittmer family on Charles Island. We also recall with gratitude the aid given us by Miguel Castro, guide, sailor, and camper *par excellence*.

Friendly support came from many of our nongeological colleagues on the expedition. In particular, we wish to acknowledge our indebtedness to Professor Charles Rick for supplying us with several specimens and photographs, to Professors Clarence Palmer, R. C. Stebbins, and Ira Wiggins for lively and profitable discussions of how animals and plants may have reached the Galápagos Islands, and to David Cavagnaro for additional photographs taken on Indefatigable and Narborough islands.

Professor G. Arrhenius supplied samples from two points on the East Pacific Rise, and Professor M. N. Bass sent samples from the Clipperton Fracture Zone and Malpelo Island. Professor Charles H. Behre, Jr., and Professor Adrian Richards loaned us aerial photographs and topographic maps of some of the islands, and the Ecuadorian Government gave consent to publish some of the pictures taken by the U.S. Air Force. Dr. Thomas Chase provided valuable bathymetric charts of the region; Dr. Brent Dalrymple supplied potassium-argon age determinations on some of the lavas; Dr. Azuma Iijima examined our palagonite tuffs and their zeolites; Professor Adolf Pabst determined the cell constants of magnetites; Dr. Alan Stueber determined the strontium isotopic composition and potassium-rubidium ratios of two samples from Charles Island and, for comparison, two from Hualalai Volcano, Hawaii; and Dr. S. R. Taylor made trace-element determinations on a group of our Galápagos rocks. To all of these gentlemen we offer warm thanks for their valuable help.

In conclusion, we offer special thanks to Dr. Ken-ichiro Aoki of Tohoku University, Sendai, Japan, for the many first-class rock and mineral analyses that form an indispensable part of our paper and for many stimulating discussions concerning them. His broad knowledge of the field of alkaline igneous rocks and their characteristic clinopyroxenes and ultramafic inclusions was of very great assistance in our studies.

Contents

TABLE

Abstract

The Galápagos Archipelago constitutes one of the world's largest and most active groups of oceanic volcanoes. The oldest rocks are on Barrington, Hood, and Indefatigable islands; they are uplifted submarine lavas of Pleistocene age. Recent volcanoes fall into three general geologic and petrographic groups. The large western islands are built of fluid tholeiitic lavas erupted from the summit and flanks of shield volcanoes, and each volcano has a summit caldera. The smaller islands along the northeastern side of the archipelago have a more complex history; they show at least two periods of activity and the rocks include both tholeiitic and magnesium-poor alkali basalts that commonly contain extremely abundant plagioclase phenocrysts. The southernmost of the central islands, including the western part of Chatham Island, are built mainly of magnesium-rich alkali basalts that form large, mature volcanoes. None of the islands in this group shows clear evidence of having developed a caldera, although parasitic vents are common. The north-central islands — James, Duncan, and Jervis — have complex histories and contain the greatest variety of rock types, including the most strongly differentiated lavas and accidental plutonic blocks found in the archipelago.

Chemically the rocks differ from those of Hawaii; they are probably typical of igneous rocks related to the East Pacific Rise. Plutonic and effusive tholeiitic rocks follow parallel courses of differentiation with enrichment in Fe/Mg, total alkalis, and excess silica. Alkali basalts are derived from a deeper source than are the tholeiitic rocks and show little evidence of differentiation.

The islands were never connected to the mainland of South or Central America. They grew from a broad shallow platform near the crest of the East Pacific Rise, and the location of individual volcanoes appears to have been controlled by two major fracture systems, one trending north-northwest and the other nearly east-west.

Review of Literature

DARWIN'S CONTRIBUTIONS

Soon after sailing from England on the *Beagle* in 1831, at the age of 22, Darwin wrote to his father from Teneriffe: "Geologizing in a volcanic country is most delightful; besides the interest attached to itself, it leads you into most beautiful and retired spots." Throughout the remainder of his long voyage, geology seems to have engrossed most of his time and attention. It was only after his return to England, as his friend Professor Judd recalled, that "biological speculations gradually began to exercise a more exclusive sway over Darwin's mind." It is proper, therefore, that tribute be paid first to Darwin for the keen and careful geological observations he made during his 5-week stay in the Galápagos Islands in 1835, especially because these and equally penetrating observations he made on the mainland of South America have generally been obscured by his better known contributions to biology.

Darwin landed first on Hood (Española) Island and then on Chatham (San Cristobal). On Chatham Island, he saw craters "composed of a singular kind of tuff" made up of yellowish-brown, translucent, resinlike fragments. He found similar craters along and close to the coasts of several other islands, and they intrigued him so much that he came to regard them as "the most striking feature of this Archipelago." His conclusion was that "much the greater part of the tuff has originated from the trituration of the grey, basaltic lava in the mouths of craters standing in the sea." Debate still continues as to how much of this kind of tuff (now called palagonite) is produced by phreatomagnetic eruptions, when hot basaltic magma rapidly absorbs water, and how much is produced during much slower absorption of water by cold basaltic glass, long after deposition, during the process of weathering. Darwin, however, was probably the first to suggest the role of sea water in the formation of some palagonites.

3

When Darwin visited the Galápagos Islands, geologists supposed that lavas containing many large crystals and vesicles must have been extremely viscous; what he saw on Albemarle (Isabela) Island clearly contradicted this opinion, indicating that other factors were involved in determining the degree of viscosity. Geologists had already speculated upon the idea that the constituent minerals of lavas might be separated according to their specific gravity and had suggested that this process might lead to extrusion of very different kinds of material, even from the same vents. Darwin seems to have been the first to substantiate this view by observations in the field. Noting the concentration of large crystals of feldspar in a thick lava flow bordering Buccaneer (Fresh Water) Bay on James (San Salvador) Island, he correctly deduced that flotation of light crystals and sinking of heavy ones might account for many variations in composition to be observed in lavas, explaining, for example, why flows of trachyte sometimes issue from a vent before flows of basalt.

When Darwin examined the lavas of the Galápagos Islands, the polarizing microscope had not been invented; nevertheless, with only a hand lens and a reflecting goniometer, he made several noteworthy observations. All of the Galápagos lavas, he said, are basalts, without "a single specimen of true trachyte." Those of the northern islands, he added, generally contain more albite (plagioclase) than do those of the southern islands, and only in one lava, from James Island, did he find crystals that "cleaved in the direction of orthite or potash felspar." He detected olivine in many flows, but never hornblende or augite.

A small cone adjoining Buccaneer Bay on James Island especially attracted his attention because the lavas there contain many angular, "pseudo-extraneous" fragments of coarse-grained material resembling syenite. It seemed to him that the lavas themselves had been formed by melting of such fragments; today, however, coarse-grained clots of this kind are regarded as "cumulates," products of early crystallization from the magmas that gave rise to the enclosing lavas; they were probably torn from the walls of the feeding pipes of the volcanoes or from the roofs of the underlying reservoirs.

Darwin was struck by the fact that in many volcanic archipelagoes the islands are arranged in arcs. He noted that in the Galápagos Archipelago, however, the volcanoes tend to be arranged in a crudely rectilinear pattern, one line trending north-northwest and the other east-northeast. He suggested that the principal volcanoes lie at intersections of these two sets of fundamental fractures. He ventured no opinion as to the cause of the rectilinear pattern; in fact, nobody has yet advanced a satisfactory explanation.

Finally, Darwin decided that proofs of uplift of the islands are "scanty and imperfect." He discovered blocks of lava cemented by shell-bearing calcareous material on Chatham Island, but they were only a few yards above high watermark. An officer told him about shell fragments, almost forming a layer in the walls of a crater, several hundreds of feet above the sea, but Darwin regarded such occurrences as of no more than local significance. Nor was he impressed by any signs of submergence, his conclusion being that the islands have always been isolated from each other and from the continent.

LATER CONTRIBUTIONS

It is surprising that so little geological work has been done in the islands since Darwin's visit, particularly because of the important role they played in the development of his concepts of the origin of species. More than 40 years passed before Teodoro Wolf, government geologist of Ecuador, made the next contributions to our knowledge of the archipelago. Subsequently, during the 1905-1906 expedition of the California Academy of Sciences, W. H. Ochsner discovered fossiliferous beds on several islands, but it was not until 1924 and 1928 that W. H. Dall published an account of the fossils, many of which he considered to be of Pliocene age.

L. J. Chubb visited the islands in 1924, as a member of the St. George Expedition from England, and in 1933 published a summary of his field observations; at the same time, Constance Richardson presented the first detailed account of the petrography of the volcanic rocks, based on the collections of both Chubb and Darwin. The first systematic geologic reconnaissance was not made, however, until 1953 when A. F. Banfield, C. H. Behre, Jr., and David St. Clair studied Albemarle Island, and the first systematic account of historic eruptions in the archipelago, written by A. F. Richards, did not appear until 1962. The first detailed investigation of regional bathymetry was that published by Shumway in 1954. There is obvious need for much more geological, paleontological, geophysical, and oceanographic work in and around the archipelago; the account presented here must be regarded as little more than that of a hasty reconnaissance.

Regional Setting of the Archipelago

The Galápagos archipelago, which consists of 14 main islands and several rocky islets, lies athwart the equator, roughly 600 miles off the coast of South America and an equal distance in the opposite direction from the crest of the East Pacific Rise. (Fig. 1). All of the main islands rise from the western end of the submarine Galápagos Platform, most of which lies at depths of less than 700 fathoms. Northeastward the platform slopes gradually toward the Panama Basin, but on the south and west its sides drop abruptly for 1000 fathoms or more to the deep sea floor. (Pl. 1). At its eastern end, the platform is separated by a saddle from the Carnegie Ridge that continues eastward almost to the coast of Ecuador, being separated only by the narrow Peruvian Trench and an equally narrow coastal bench.

Another important submarine feature is the Galápagos Fracture Zone, marked by high and closely spaced ridges and intervening furrows. It runs westward, approximately along latitude 2°N., from near Culpepper (Darwin) Island, at the northern end of the archipelago, until it cuts and offsets the crest of the East Pacific Rise and then continues beyond, its total length being about 1600 miles (Shumway and Chase, 1961). A third major submarine feature is the Cocos Ridge, extending northeast from near the northern end of the archipelago almost to the coast of Costa Rica, from which it is separated only by the narrow eastern end of the Middle American Trench. Further reference is made to the foregoing features on later pages, in discussing the fracture patterns of the archipelago and the possibility of former connections with the continent (p. 99).

In the notes that follow, we begin by describing the oldest rocks of the archipelago, namely the uplifted submarine lavas, tuffs, and limestones found on the south-central islands. We then describe the younger volcanoes, beginning with those that form the southern islands and proceeding northward. In describing each island in turn, we have considered it best to note the petrographic character of its rocks, leaving the general petrological summation for a later section of the paper (pages 117-180).

7

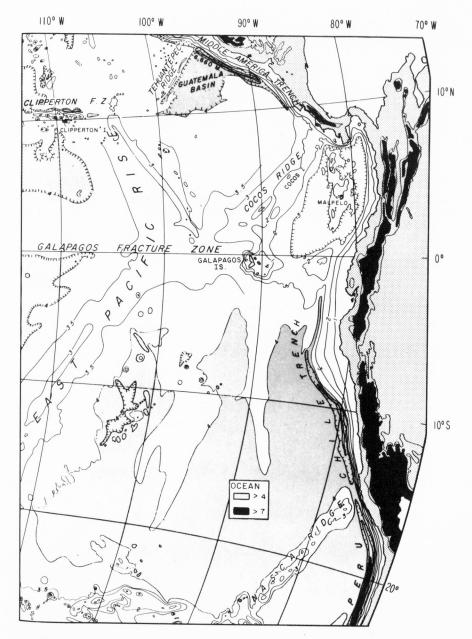

Figure 1. Bathymetry of the southeastern Pacific Ocean. Shows the relation of the Galápagos Islands to the major submarine features. Depths in kilometers. (*From* Menard, 1964.)

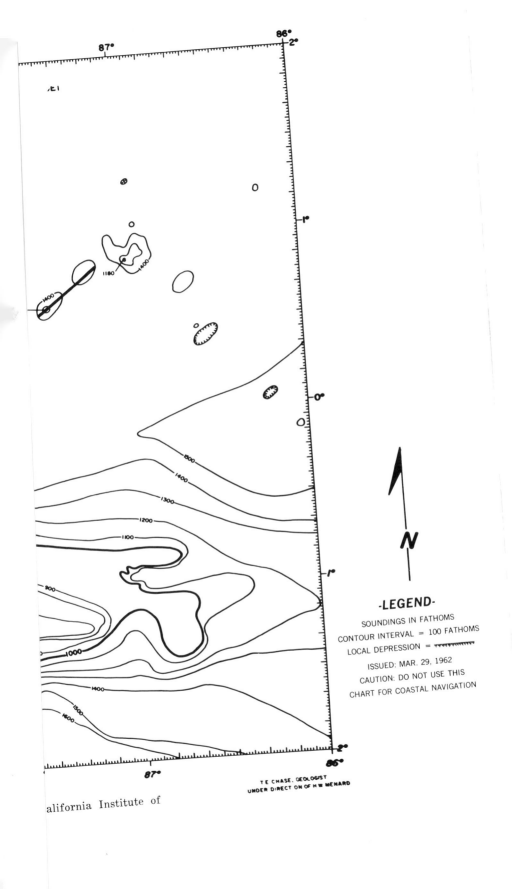

87° 86° 2°

/El

0

1°

1180 1400

1400

0°

1°

-LEGEND-

SOUNDINGS IN FATHOMS
CONTOUR INTERVAL = 100 FATHOMS
LOCAL DEPRESSION = ∿∿∿∿∿∿∿
ISSUED: MAR. 29, 1962
CAUTION: DO NOT USE THIS
CHART FOR COASTAL NAVIGATION

1500

1400

1300

1200

1100

900

1000

1400

1500

1600

2° 86°

87°

T E CHASE, GEOLOGIST
UNDER DIRECT ON OF H W MENARD

alifornia Institute of

N

Uplifted Submarine Volcanic Rocks

HOOD (ESPAÑOLA) ISLAND

No detailed geological study has ever been made of this island. This was where Darwin first went ashore in the archipelago, but his stay must have been very brief, for he wrongly affirmed that craters were to be seen there. Chubb did not visit the island, and he wrongly states that it consists of a single volcano "of a type similar to Charles Island"; Richards also refers to it erroneously as a single volcano of the "shield type." The report of an eruption on the island in 1958 is certainly mistaken.

Hood Island is definitely not a separate volcano but an uplifted block of submarine lavas, comparable in this regard to Barrington and Baltra islands. It measures approximately 10 miles in an east-west direction and about 5 miles across and rises to a maximum height of a little more than 650 feet (Fig. 2). There are no high cliffs along the northern coast, but along the opposite coast the cliffs rise gradually from the extremities of the island to heights of more than 300 feet near the middle. These formidable cliffs are only partly due to marine erosion; primarily they represent a battered fault scarp. The lavas of the island have been gently warped, upheaved, and tilted northward, forming a broad, northward-plunging anticline. A second fault, approximately parallel to the one bordering the southern coast, may traverse the island, but this is a conjecture based solely on the topography.

There are no remnants of parasitic cones anywhere on the island; the occasional conical peaks that one sees are simply erosional features cut in lavas that dip gently in northerly directions. The entire western third of the island, from Punta Suarez to the summit peak, consists of massive, dense, pale-gray flows of olivine basalt similar to those on Barrington and Baltra islands. They dip to the west and northwest at low angles, so as one goes eastward along the coastal cliffs from Punta Suarez he sees an increasing number of flows. At Punta Suarez itself, only two flows or flow units are exposed; about 3 miles to the east there are seven, and 4 miles to the east

9

there are perhaps fourteen. Individual sheets vary in thickness between approximately 10 and 60 feet, most of them averaging between
20 and 30 feet. Each has a basal and a crustal layer of reddish, rubbly,
scoriaceous material, but none was observed to have a pillow structure.
Three flows tested by Allan Cox near Punta Suarez showed reverse
magnetization, and these, it should be noted, are among the youngest
on the island.

The long and gentle slopes that rise from Punta Suarez to the
summit of the island are essentially dip slopes, and in many places
they are studded with small, mesalike tors or monadnocks of massive
lava that stand a few feet to a few tens of feet above the general
surface, most of which is cut in reddish, scoriaceous basalt. Seen from
a distance these curious erosion forms look like flat-topped buildings.
They appear to have been etched into relief by marine erosion, once
forming sea stacks rising from a wave-cut platform eroded mainly in
softer, scoriaceous lavas. Some support is given to this interpretation
by the presence of a few pebbles and cobbles of limestone, such as
those found by Theodore Papenfuss near the summit peak. We saw

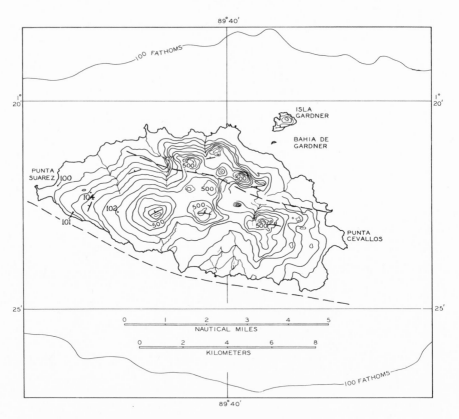

Figure 2. Map of Hood (Española) Island. Contour interval 50 feet. (*After*
U.S. Hydrographic Office chart No. 5944, 1st ed., 1947.)

no marine limestones interbedded with the lavas near Punta Suarez, but they may well be present in other parts of the coastal cliffs.

Petrography

Chesterman (1963) described an olivine basalt from Hood Island, noting particularly that it contains purple titaniferous augite. Most of the lavas that we examined microscopically are alkali-olivine basalts of very uniform character. The only phenocrysts they contain are small but abundant ones of olivine ($2V = \sim 90°$), many of which are partly altered to iddingsite. The plagioclase, accounting for 35 to 40 percent of the total volume, is mostly medium or calcic labradorite. In some lavas it forms glomerophyric clusters, along with olivine and augite, but more commonly it occurs as zoned laths up to about 1 mm in length. Small amounts of anorthoclase are present between some of the laths. Most of the clinopyroxenes are pale-reddish-brown to purplish-brown, ophitic augites, presumably somewhat titaniferous; they constitute about one-third of the volume. Ilmenite and titaniferous magnetite are unusually abundant, commonly exceeding 10 percent. Analcite is a minor but constant accessory, and in some specimens a zeolite, probably natrolite, amounts to 1 or 2 percent. An analysis of a typical lava is presented in Table 1 (No. 102); this specimen, which is illustrated in Figure 3c, came from the line of crags at an elevation of about 400 feet, some 2.5 miles east of Punta Suarez.

A specimen collected from the summit peak by Theodore Papenfuss differs from all others examined, both in texture and mineral composition. It is a zeolite-rich basalt containing little if any olivine. About one-half of the rock consists of calcic labradorite; intergranular, greenish augite accounts for one-third. Granules of magnetite and subordinate hematite together make up 10 percent. At least two varieties of zeolite are present, one as concentric layers lining small cavities and the other as round patches, apparently replacements of earlier minerals. Calcite fills many small vesicles. The largest individual grain seen in thin section was a euhedral crystal of clear brown spinel, about 0.17 mm across, with a narrow, opaque reaction rim.

BARRINGTON (SANTA FE) ISLAND

Barrington Island is a roughly rectangular horst about 5 miles long and 3 miles wide (Fig. 4). It consists of a group of gently tilted and flat-lying fault blocks trending approximately east-west and composed mainly of basaltic lavas uplifted from the sea. Some of the peaks on the island exceed 800 feet in height, and the coast line is

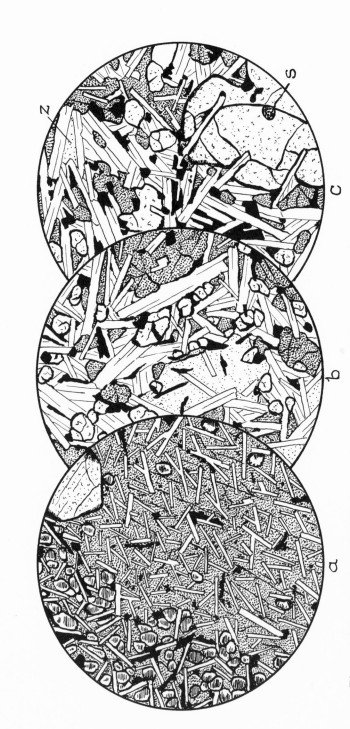

Figure 3. Microdrawings of uplifted submarine basalts. (a.) Barrington Island (No. 111). Part of the lava consists of large poikilitic plates of augite, up to 2 mm across, enclosing laths of labradorite and a little ore; part is intergranular, the feldspars being separated by granules of olivine and fewer of augite, along with abundant ore. (b.) Baltra Island (No. 31). Labradorite laths, olivine, purplish titaniferous augite, ilmenite, magnetite, and (shown by stipples) analcite. (c.) Hood Island (No. 102). Olivine, purplish augite, labradorite laths, ilmenite, and titaniferous magnetite. Small amount of interstitial anorthoclase and accessory zeolites (Z). Grain of brown spinel (S) is included in the largest olivine crystal. Analysis is given in Table 1. This alkali-olivine basalt is essentially identical to No. 103 from the main volcano of Chatham Island. Diameter of each field is 2 mm.

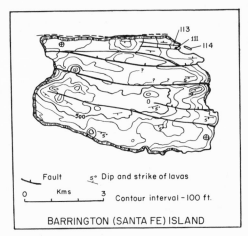

Fault   5° Dip and strike of lavas

0   Kms   3

Contour interval - 100 ft.

BARRINGTON (SANTA FE) ISLAND

Figure 4. Map of Barrington (Santa Fe) Island. (Base *after* U.S. Hydrographic Office chart.)

marked by impressive cliffs. Nowhere are there any signs of recent volcanism; constructional slopes are totally lacking, the landforms being entirely erosional in origin.

The subparallel fault blocks that comprise the Barrington horst vary in width from less than a quarter of a mile to about a mile, and, although some are essentially horizontal, most are tilted southward at angles of 5° or less. High cliffs along the northern coast coincide closely with a fault plane; indeed, at one locality either the main fault plane or a parallel, subsidiary one is well-exposed, separating a thin sliver that dips steeply northward from an adjacent block that dips in the opposite direction. The cliffs along the southern coast must also represent a battered fault scarp.

Almost all of the fault scarps that cross the island are conspicuous features, and some are remarkably fresh, indicating movements within the last few thousand years. One such youthful, almost precipitous scarp is to be seen near the northeastern end of the island, where it trends directly toward one of the youthful fault scarps bordering Academy Bay on neighboring Indefatigable Island. The uplifts that produced Barrington, Baltra, and Hood islands may well have taken place long before the subaerial volcanoes of the archipelago began to grow, but it is clear that as these volcanoes grew the uplifted, submarine lavas continued to be displaced by faulting.

Our own visit to Barrington Island lasted only a few hours and was confined to the northeastern corner, close to the usual landing place. Fortunately, Professor Wyatt Durham was able to spend much more time on the island; we are therefore grateful to him both for the loan of specimens and for an account of his observations.

Close to the usual landing place there is a small islet; here Professor Durham discovered a thin lens of fossiliferous marine limestone interbedded with olivine-rich basalts. About a quarter of a mile inland, due west of the islet, he collected another olivine-rich basalt with a porous, diktytaxitic groundmass, very much like the oldest of the uplifted submarine lavas near Cerro Colorado, on the northeastern coast of Indefatigable Island. He also noted palagonitic lapilli tuffs and tuff breccias near the northwestern end of Barrington Island similar to those near Cerro Colorado. And, while at Academy Bay, he received from Mr. Dubois, a local resident, a beautiful, zoned crystal of augite, about half an inch long, collected near the eastern end of Barrington Island, probably from a bed of palagonite tuff between the lava flows. Large crystals of augite are also present in the tuffs near Cerro Colorado, but, as far as we know, they are entirely absent from all the lavas of the young volcanoes of the archipelago. These observations, taken together, suggest that the lavas and tuffs of Barrington Island may be approximately coeval with the uplifted submarine lavas on Indefatigable and Baltra islands. The fact that two lavas examined by Dr. Allan Cox from the northeastern end of Barrington Island showed normal magnetization does not invalidate this view.

The basalts exposed on the steep fault scarp close to the landing place are dense, dark-gray lavas with abundant phenocrysts of olivine but none of feldspar. On fresh surfaces they appear to be quite structureless, but weathering reveals a pronounced, contorted fluidal banding. Ten or twelve flows are to be seen on the scarp, some of them up to 40 feet thick although most vary between 5 and 20 feet in thickness. Each has an autobrecciated top and bottom, and in places they are separated by lenses of pyroclastic debris. We think that all of these rocks were laid down in shallow seas before being raised to their present positions.

Petrography

The lavas that we examined from Barrington Island have an unusual texture that they share with some of the oldest lavas of Indefatigable and Baltra islands. Phenocrysts of olivine ($2V_x = 89°$) amount to as much as 5 percent of a typical sample; microphenocrysts of calcic labradorite, commonly enclosed within glomerocrysts of olivine, make up less than 1 percent. These larger crystals lie in a groundmass composed of sodic labradorite laths, olivine, and a mosaic of clinopyroxene oikocrysts. The plagioclase laths, which range up to 0.5 mm in length, are distributed regularly but are oriented at random, and most are enclosed ophitically in irregular patches of clinopyroxene up to 2 mm across. Outside the areas of clinopyroxene, the laths of plagioclase are accompanied by intergranular augite and olivine (Fig. 3a).

Professor Hisashi Kuno pointed out to us that this unusual texture resembles that of many high-alumina basalts in other parts of the world. The Barrington basalt analyzed by Dr. Aoki (No. 111, Table 1) does in fact lie within the field designated by Professor Kuno for high-alumina lavas of the same silica content (Kuno, 1960, p. 130). The chemical character of the lavas and of the large augite phenocryst given to us by Professor Wyatt Durham are discussed in more detail on pages 117-119.

INDEFATIGABLE (SANTA CRUZ) ISLAND

Indefatigable Island consists of two distinctly different parts; the younger and by far the larger part is a broad shield of basaltic lavas, roughly oval in plan, measuring about 25 miles in an east-west direction and 20 miles north-south, surmounted by a cluster of youthful scoria cones; the older part, revealed in a narrow strip along the northeastern coast and perhaps also at the northwestern corner of the island, consists of uplifted submarine lava flows and tuffs interbedded with fossiliferous limestones. Our concern here is with the older part of the island only; the younger part is described on pages 31 to 35. We realize, in retrospect, that much of our time should have been spent on a study of the northeastern coast of the island, between Cerro Colorado and the Itabaca Channel, for the cliffs along this 10-mile stretch undoubtedly reveal the finest geological section anywhere in the archipelago. Our information concerning this important section comes almost entirely from observations made by our colleague, Professor Wyatt Durham, to whom we are therefore greatly indebted.

The coastal cliffs extending northward from Cerro Colorado reveal a sequence of north-dipping, uplifted basaltic flows and tuffs and interbedded marine sedimentary rocks, the aggregate thickness of which must be very great, even though the section is partly repeated by strike faults. One such fault forms a small scarp immediately north of Cerro Colorado, and probably continues out to sea, passing through the channel between the two Plaza Islands (Fig. 5). These islands are themselves fault blocks composed of uplifted submarine flows and sediments, the southern block being tilted slightly to the north and the other lying horizontal. A sample collected by Dr. Edgar Martin from the western end of South Plaza Island is a "big feldspar basalt."

The oldest of the uplifted rocks close to Cerro Colorado are lapilli tuffs and tuff breccias. They contain many angular and subangular fragments of both dense, nonporphyritic basalt and "big feldspar basalt," some of which are amygdaloidal, along with fragments of fossiliferous limestone. These ejecta lie in a brownish matrix of sideromelane and palagonite tuff and tuffaceous sand, commonly rich in zeolites. Large crystals of augite are plentiful in the tuff, similar to those already noted from tuffs on Barrington Island.

Above the basal deposits lies a series of fossiliferous limestones, basalts, palagonite tuffs, and tuffaceous sediments. Some basalts show pillow structures, and at least one is associated with palagonitic debris that exhibits fore-set bedding, almost surely a sign of subaqueous deposition. At a nearby locality, however, Professor Durham observed a basaltic lava that had reddened and baked the tuffaceous material over which it spread, indicating that it probably flowed over dry land. He also noted (1965, p. 6) "on a small isolated fault block,

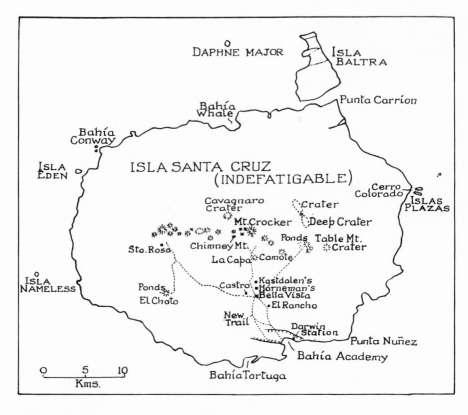

Figure 5. Sketch map of Indefatigable (Santa Cruz) Island. Based in part on observations by David Cavagnaro.

near Cerro Colorado, a few feet of indurated, crossbedded calcareous sandstone suggestive of subaerial deposition."

The evidence gathered by Professor Durham leaves no doubt that, although most of the uplifted and tilted rocks near Cerro Colorado were laid down in shallow seas, almost surely some of the volcanic rocks accumulated on land, perhaps on low islands. Subsidence below sea level must therefore have taken place before the entire sequence was raised and tilted to its present position. Uplifted lavas and palagonite tuffs may also be present, according to Professor Durham, at the northwestern end of Indefatigable Island, near Conway Bay.

Petrography

Two specimens of the uplifted submarine lavas from the coastal cliffs a short distance north of Cerro Colorado are nonporphyritic basalts. They consist of small laths of medium labradorite in a groundmass that has a texture common in many rapidly quenched basalts.

Subradiating slender prisms of clinopyroxene, with fine opaque dust along their cleavages, are surrounded by dark-brown glass; amygdules are filled with calcite and zeolites.

Among the palagonite tuffs associated with these submarine flows are many lithic blocks of strongly porphyritic basalt (Fig 6). Some of these contain as much as 30 percent phenocrysts of clear, medium labradorite with rims of labradorite-andesine. Microphenocrysts of olivine and subcalcic augite are quite subordinate. The hyalopilitic groundmass is made up of sodic labradorite, clinopyroxene, reddish iddingsite pseudomorphs after olivine, and magnetite-laden glass. Vesicles and fractures are lined with calcite and fibrous zeolites.

BALTRA AND SEYMOUR ISLANDS

North of Indefatigable Island, separated from it only by the narrow Itabaca Channel, lies the low island of Baltra, and to the north of that, separated from it by the narrow Northern Channel (Canal del Norte), is the much smaller island called Seymour. These two islands consist of a series of fault blocks trending approximately east-northeast, as indicated in Figure 7. They are composed of basaltic flows and interbedded limestones laid down beneath the sea and later uplifted.

On Indefatigable Island, as noted already, all the uplifted lavas dip northward; on Baltra and Seymour islands they dip both north and south. Thus the Northern Channel occupies a graben between outward-dipping blocks, whereas Itabaca Channel occupies a graben between blocks that dip inward. The graben in the center of Baltra Island, part of which is occupied by Aeolian Cove, lies between outwardly dipping blocks. It must be emphasized, however, that the dips of all these tilted blocks are very low, nowhere exceeding about 5°. Their east-northeast trend is approximately at right angles to the large submarine fault scarp to the east of the islands.

Ochsner, in 1906, was the first to discover fossiliferous limestones between the lavas on Baltra Island. Two of the fossil localities he found adjoin Aeolian Cove; the third, which we did not visit, lies on the eastern coast, about 2 miles to the northeast. Dall and Ochsner (1928), who published their joint studies 22 years later, regarded the fossils as of Pliocene age. One of Ochsner's discoveries was made in the small cliffs near the landing place at the northeastern corner of Aeolian Cove, where roadcuts now reveal a bed of limestone, locally almost a coquina, about 3 feet thick, between flows of basalt. Volcanic debris is almost completely lacking in the limestone; indeed, nowhere on Baltra Island did we see tuffaceous sediments between the lavas. Remnants of what appears to be the same bed of limestone cap the southward-tilted fault block south of Aeolian Cove, and an older bed

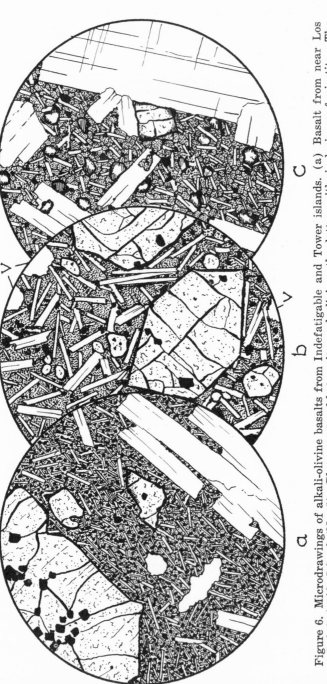

Figure 6. Microdrawings of alkali-olivine basalts from Indefatigable and Tower islands. (a) Basalt from near Los Pozos, Indefatigable Island (No. 3). Phenocrysts of bytownite and olivine, the latter with inclusions of picotite. The groundmass consists of labradorite laths, with granules of titaniferous augite and euhedral magnetite; elsewhere in the groundmass, but not shown, are clusters of picotite granules. (b) Basalt from Darwin Station, Indefatigable Island. Analysis in Table 2c, No. 1. Phenocrysts of olivine, with inclusions of picotite, accompanied by laths of labradorite, small granules and subophitic plates of titaniferous augite, and magnetite and ilmenite. Not recognizable in the drawing are a little interstitial brown glass and anorthoclase and minute needles of apatite. V = vesicles. (c) Porphyritic basalt from Tower Island (No. 92). Collected by Professor J. C. Granja. Large phenocrysts of bytownite, and a smaller one of olivine, lie in an intergranular matrix of labradorite laths, granules of augite, olivine (partly altered to iddingsite), and titaniferous magnetite. Diameter of each field is 2mm.

of shell-rich limestone is to be seen in the cliffs on this side of the cove, immediately beneath the topmost lava.

All of the lavas are extremely massive, with strong, broadly spaced joints; weathering therefore produces piles of huge, angular and subangular boulders. A short distance north of the landing place in Aeolian Cove, narrow clefts in the lavas, some of them more than 50 feet deep, mark tension features along the crest and face of a fault scarp.

Two lavas bordering Aeolian Cove were found by Dr. Allan Cox to exhibit reversed magnetism, and one of these was found by Dr. Brent Dalrymple of the U.S. Geological Survey to be approximately 1.47 m.y. old.

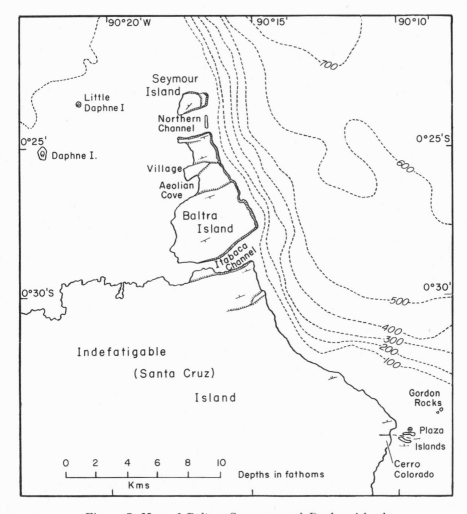

Figure 7. Map of Baltra, Seymour, and Daphne islands.

Petrography

Specimens from six successive flows, including the oldest on the island, were examined microscopically. The lowest flow in the sequence (No. 28) is a medium-grained basalt of seriate texture, about half of which consists of normally zoned crystals with cores of bytownite-labradorite and rims of andesine. These crystals range in size from 3 mm down to microlites a few tenths of a millimeter in length. In addition, there are irregular interstitial grains of andesine. Augite ($2V_z = 55°$) constitutes about one-third of the total volume, and its size range is almost the same as that of the plagioclase. Most of the augite is intergranular, but a few large grains have grown ophitically around the plagioclase. Olivine ($2V_x = 88°$) amounts to about 5 percent of the volume, most of it being partly changed to iddingsite. Ilmenite, magnetite, hematite, anorthoclase, accessory apatite, and interstitial glass account for the remainder. A chemical analysis is presented in Table 1; it indicates more than 1 percent of normative quartz.

The next two lavas in the sequence are essentially alike; both are analcite-bearing olivine basalts consisting of rare microphenocrysts of augite and labradorite in a fine-grained intergranular groundmass of labradorite, clinopyroxene, and granular ore. A chemical analysis of the lower of the two lavas (No. 29) is given in Table 1.

The next higher flow is a coarse-grained, analcite-bearing alkali-olivine basalt (Fig. 3b). About half of it consists of laths of sodic labradorite enclosed ophitically by purplish titanaugite ($2V_z = 55°$) that forms about one-third of the lava. Olivine (2V close to 90°) accounts for about 10 percent, and it contains many inclusions of picotite. Ilmenite and magnetite constitute approximately 3 percent, accessory zeolites, calcite, apatite, a little sodic augite, and glass making up the remainder. Chemical analysis (No. 31, Table 1) shows that the rock is distinctly more alkaline and poorer in silica than the older lava (No. 28).

The two uppermost lavas are strongly porphyritic, with phenocrysts of medium labradorite comprising about one-third of their volume. Zoned phenocrysts of augite are smaller and make up only about 5 percent. Phenocrysts of olivine, most of which have reddened rims, are about equal to those of augite in size and amount. The groundmass consists of labradorite (~50 percent), augite (~35 percent), olivine (~10 percent), ilmenite, magnetite, and interstitial glass.

Younger Volcanoes

CHARLES (FLOREANA OR SANTA MARÍA) ISLAND

No island in the archipelago is more attractive scenically, particularly when viewed from a distance, than Charles Island. Its gentle, green slopes rise to uplands generally between 1000 and 1300 feet in elevation. Richards (1962) thought that the island was "a single shield volcano with an eroded summit caldera" 5 miles wide, and Chubb (1933) said it was similar in structure to the large volcanoes of Albemarle and Narborough islands but that its summit crater was "reduced to a circle of hills surrounding a fertile plateau." The map (Fig. 8) clearly shows, however, that there is no central caldera. If there ever was one, and this is not unlikely, it has been completely obliterated by the products of later eruptions. To use the analogy of Hawaiian volcanoes, the Charles Island volcano may be in the "Mauna Kea stage" of development, an earlier shield volcano with a summit caldera comparable to those of Mauna Loa and Kilauea having been buried by swarms of younger cones and flows. If this is so, then the Charles Island volcano has much in common with the major volcanoes of Bindloe and Indefatigable islands. No other major volcano in the archipelago is more densely crowded with parasitic cones; indeed, there are more than 50 within an area of approximately 80 square miles. There seems to be no systematic arrangement of these cones, except for a north-northeast alignment of three near Daylight Point: certainly there is no circular pattern of cones such as might reflect the outlines of a buried caldera.

Our stay on the island was limited to five days. We examined the coastal belt from Rada Black Beach northward to Daylight Point and thence eastward to Post Office Bay and Cormorant Point; we followed the trail to the old Wittmer home at the foot of Mount Olympus (Wittmerberg). Most of the rest of the island was seen from afar while skirting the shore by boat.

Chubb thought that Charles Island should be grouped with Chatham and Hood islands as probably the oldest in the archipelago,

21

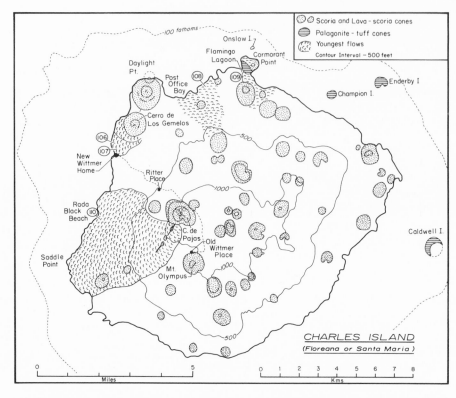

Figure 8. Map of Charles (Floreana or Santa María) Island. (Topographic base *after* U.S. Hydrographic Office chart No. 5940, 1st ed., 1946.)

and Banfield and his colleagues likewise thought that this south-eastern part of the archipelago was the oldest. Our observations do not support these views. No geologist, seeing Charles Island from a distance or setting foot on shore, would hesitate to say that almost all of the visible lavas and cones, with the exception of a few flows in the coastal cliffs, are of very recent origin. Over wide areas, sheets of black, pahoehoe lava, several of them almost barren of vegetation, preserve minute details of their surface features. Some lavas in the cliffs south of Daylight Point have been shown by Allan Cox to have reversed magnetism and therefore to be more than about 0.7 m.y. old, and additional work may reveal similar lavas in cliffs along the southern coast; otherwise all the visible lavas seem to have been discharged more recently, and almost surely some were discharged within the last thousand years, if not within the last few centuries. Captain Porter of the U.S.S. *Essex* reported that in July 1813 he saw "a volcano burst out with great fury" from the center of the island (Richards, 1962, p. 30), and although we found neither lavas nor fragmental ejecta of such recency in the areas we traversed, they may well be present in the uplands east of Cerro de Pajas. Other writers may have been misled into thinking that Charles Island is one of the oldest in the archipelago because of the fairly dense cover of vegetation, even at low levels. In our opinion, this relatively luxuriant growth probably reflects a greater rainfall than is normal in the archipelago, and certainly it results in part from the thick mantle of basaltic ash laid down by eruptions from parasitic cones.

By far the most conspicuous cone on the island, and perhaps the youngest, is Cerro de Pajas, which rises to 2100 feet, approximately 1100 feet above the surrounding terrain (Pl. 2). It consists chiefly of sand- and gravel-sized basaltic ejecta mingled with angular lapilli and blocks of dense basalt. Rounded bombs and clinkery, scoriaceous fragments are notably scarce; in fact, most of the ejecta must have been solid or almost so when erupted. There is no summit crater, but there is a large breach on the southern flank of the cone, from which a long line of agglutinate and scoria conelets extends southwestward. Voluminous flows of pahoehoe lava poured from these vents, inundating more than 12 square miles and reaching the coast over a front of more than 4 miles, from south of Saddle Point to north of Rada Black Beach. These flows are completely barren in places, and elsewhere are only thinly mantled by *Bursera* forest. They were discharged after explosive activity had come to an end at Cerro de Pajas, and they may well be the last flows to be erupted on the island. They consist of alkali-olivine basalt. Huge boulders of the lava are piled high on Rada Black Beach, and among these are sporadic, angular inclusions of dunite and peridotite, up to 4 inches or so across, descriptions of which are given below.

About half a mile west of Cerro de Pajas, not far from the old Ritter house (now the Kruz farm), is a slightly older cone, approximately 200 feet high and a little more than half a mile across the base. Part of the rim of its shallow summit crater is formed by a steep-sided pile of coarse spatter or agglutinate, made up of flattish cakes of scoriaceous basalt, 1 to 2 feet thick, that were sufficiently plastic when they fell from the air to adhere to one another. The remainder of the cone is mantled with ejecta of sand and gravel size, but whether these are products of the last eruptions of the cone itself or were blown from Cerro de Pajas was not determined.

The mantle of ejecta overlying much of the western part of the island contains few fragments more than an inch across; most of them are less than half that size. Some fragments are vesicular enough to be called "cinders," but most are angular chips of dense basalt, presumably blasted from the congealed fillings of volcanic conduits or from the walls of craters.

Some of the ash- and cinder-covered lavas south of the Wittmer trail, and extending southward along the coast for about a mile from

VOLCANIC PROFILES

Part of Charles Island, with Rada Black Beach in the foreground. Three large scoria cones rise above the gentle lava slopes, the highest being Cerro de Pajas. Mount Olympus (Wittmerberg) on the right.

McBIRNEY AND WILLIAMS, PLATE 2
Geological Society of America Memoir 118

the new Wittmer place, seem to have been erupted from the base of the Ritter Çone. Their cover of mesquite contrasts vividly with the *Bursera* forest on the younger lavas from Cerro de Pajas.

Mount Olympus, a mile and a half south of Cerro de Pajas, is a craterless cone of well-bedded basaltic cinders. Most of the ejecta are moderately compacted, and in places they exhibit cavernous weathering. One of the few good springs on the island issues from the northern foot of the cone, close to the artificially modified caves where the Wittmers made their first home.

Immediately north of Rada Black Beach is an almost barren flow of aa and scoriaceous basalt devoid of phenocrysts of feldspar but containing many of olivine. Locally, it carries clots, up to 6 inches across, composed of small olivine crystals accompanied by a few of chromite. A slightly older and more vegetated flow separates this basalt from the lava-scoria cone of Cerro de los Gemelos.

Along the coast west of Cerro de los Gemelos are cliffs up to 40 feet high cut in columnar basalts, some of which, according to Allan Cox, carry abundant inclusions of dunite as well as large phenocrysts of plagioclase, a combination rarely seen in any other lavas on the island.

North-northeast of Cerro de los Gemelos, on the same eruptive fissure, is another lava-scoria cone, the coastal section of which reveals many flow units 1 to 2 feet thick; beyond this cone, on the same fissure, lies a third cone, now bisected by marine erosion to form Daylight Point. The activity that produced these three cones appears to have begun at the northern end of the fissure and to have migrated southward.

Most of the area adjoining Post Office Bay is occupied by extremely youthful flows of pahoehoe lava characterized by abundant pressure mounds. These flows have only a scant cover of vegetation and none of cinders. Like almost all other lavas on the island, they lack phenocrysts of plagioclase but are lightly sprinkled with small ones of olivine. Their source seems to be a large cone of reddish scoria about 2 miles inland.

Cormorant Point has two distinct parts, an older, western part composed of thinly bedded, yellowish-brown sideromelane and palagonite tuff, and a younger, eastern part composed of black basaltic lava and scoria. Half of the tuff cone had already disappeared, more likely by downfaulting than by marine erosion, before the lava-scoria cone began to erupt. Short tongues of basalt issued from the younger cone, one of them spreading along the northern side of the tuff cone and extending out to sea to form Onslow Island.

Less than a mile to the south of these cones is a much larger one of brown and black basaltic scoria, and at least three other cones rise a short distance to the east. Adjoining these cones are widespread sheets of very youthful, scoriaceous and blocky basalts, some of which descend to the shores of Flamingo Lagoon.

The small islands off the western coast of Charles Island, namely Champion, Enderby, Caldwell, and Gardner, appear from a distance to be eroded cones of well-bedded palagonite tuff.

Petrography

The lavas of Charles Island are alkali-olivine basalts; all contain phenocrysts of olivine, but most are devoid of phenocrysts of feldspar. The specimens we examined under the microscope vary only in the crystallinity of the groundmass and the proportion of olivine phenocrysts. Many lavas contain abundant ultramafic inclusions, and at least some of the olivine and pyroxene phenocrysts in the lavas were derived by their dismemberment.

A specimen from the very recent lava south of Cerro de los Gemelos is typical. It contains large phenocrysts of olivine (Fa_{32}), many of which are zoned and spotted with octahedra of picotite. Phenocrysts of greenish diopsidic augite ($2V_z = 56°$) are less numerous. A few fragments of plagioclase are clearly xenocrysts derived from ultramafic inclusions; they are mainly broken crystals of labradorite with zones of glass inclusions and clear sodic rims. The groundmass of the lava is made up of slender laths of calcic labradorite, some of which are separated by patches of anorthoclase. Purplish-brown augite ($2V_z = 51.5°$), fine-grained olivine, magnetite, and a little interstitial glass make up most of the remainder. A small amount of analcite can be found in nearly every cavity. Specimen 108, from a flow immediately inland from Post Office Bay, is a similar basalt but of coarser texture and with many patches of zeolites.

The ultramafic inclusions differ widely in bulk composition, forming a continuous series from dunites with only 1 or 2 percent spinel and little if any pyroxene (for example, No. 110, Fig. 9c) to wehrlites consisting of chromian diopside, olivine, and hercynite. Some wherlites (for example, No. 99, Fig. 9a) contain a little interstitial labradorite, probably the source of the feldspar xenocrysts in the enclosing lavas. In contrast to the wide variation in the bulk composition of the inclusions, their constituent minerals are remarkably uniform. The composition of a typical clinopyroxene is given in Table 16, No. 98; the bulk composition of the total rock from which the pyroxene was separated is given in Tables 5, 6 and 7, No. 98. Both the inclusion and the enclosing lava are illustrated in Figure 9b.

The optical properties of all the clinopyroxenes we examined from the ultramafic inclusions are identical within the margin of error of the measurements, but they differ markedly from those of the pyroxenes of the host lavas, these being poorer in calcium and

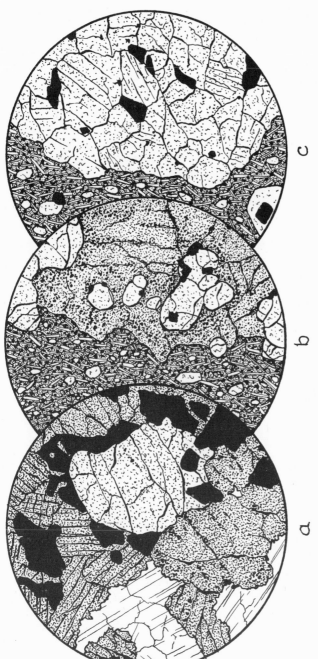

Figure 9. Microdrawings of ultramafic inclusions from Charles Island. Diameter of each field is 6 mm. (a) Labradorite-bearing wehrlite (No. 98). Green chromian diopside, olivine, and very dark green hercynite. (b) Peridotite (No. 99). Green chromian diopside and olivine with accessory magnetite and chromite. The surrounding alkali-olivine basalt has a fine-grained, intergranular texture and consists of granules of olivine, augite, and ore with laths of labradorite. (c) Dunite (No. 110), with inclusions of dark-brown spinel, enclosed in porphyritic alkali-olivine basalt. Dismembered fragments of dunite are scattered throughout the lava. Locally the lava contains a little interstitial brown glass, but mostly the texture is intergranular, laths of plagioclase being separated by granules of ore, olivine, and titaniferous augite.

richer in iron. Similarly, the olivines of the ultramafic inclusions seem to be richer in magnesia than those in the groundmass of the host basalts, but we have fewer data to support this generalization. Further observations on the ultramafic xenoliths are presented on page 130.

CHATHAM (SAN CRISTOBAL) ISLAND

Chatham Island is often referred to as one of the oldest islands of the archipelago, a view apparently based on the fact that the high western part is densely covered with vegetation. It is, unfortunately, one of the least studied islands, and our own stay of two days permitted no more than a quick trip from Wreck Bay to the uplands around Progreso and a boat trip to Sappho Cove and the spectacular Kicker Rock offshore.

The island consists of two quite distinct halves (Fig. 10). The older, western half is formed by a single large volcano that rises to a height of more than 2300 feet, with a few parasitic cones on its flanks; the eastern half, on the contrary, lies at elevations of less than 500 feet and is formed by a cluster of minor cones and attendant lavas. The boundary between these two halves may be a concealed fault, approximately in line with the profound submarine scarp west of

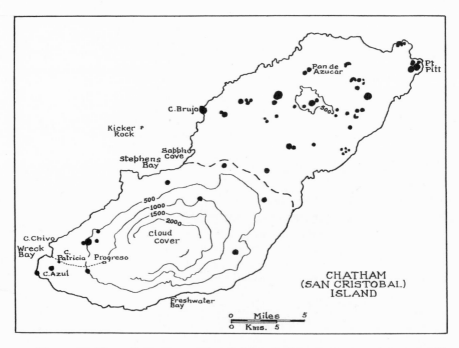

Figure 10. Map of Chatham (San Cristobal) Island. Contour interval, 500 feet. Black circles denote small lava and scoria cones. (Topographic base *after* U.S. Hydrographic Office chart.)

Abingdon and Bindloe islands and with Culpepper and Wenman islands.

The volcano occupying the western part of the island measures 17 miles in an east-west direction and 9 miles across. When aerial photographs were taken as a base for the topographic map, clouds covered the higher parts of the island; hence nothing is shown on the topographic map in the area above 2300 feet. During our visit, clouds again obscured most of the summit area. It is said, however, that there is a large cone there, containing a crater lake more than a mile across.

The main volcano is composed almost wholly of basaltic lavas. Those exposed on the rocky beach at Wreck Bay and along the road that climbs 1000 feet to Progreso are blocky flows of olivine basalt, most of which have a microporous, diktytaxitic texture. Their content of olivine phenocrysts averages about 5 percent by volume but ranges from almost zero to 15 percent; phenocrysts of plagioclase are notably absent. Cerro Patricia, adjoining the road to Progreso, about half a mile from Wreck Bay, is a steep-sided mound of similar lava, probably built over an eruptive vent; but Cerro Azul, a mile or so south of Wreck Bay, seems from a distance to be a scoria cone with a stumpy flow of barren lava on its northern flank.

Above Progreso, and for at least 2 miles eastward, no solid outcrops were observed, the only visible rocks being a few round boulders of olivine basalt in a deep, clayey soil. Weathering increases rapidly at still higher elevations, so that the cover of clay and vegetation continues to increase in thickness.

Cerro Chivo, immediately north of Wreck Bay, and Cerro Mundo, approximately 2 miles to the east, are large mounds of basaltic lava, comparable with Cerro Patricia, and all the northern flank of the main volcano is also mantled with flows of massive, blocky basalt. Judging by the slight amount of erosion, the main volcano seems to be of about the same age as the main volcanoes on James and Indefatigable islands; the youngest flows are presumably of Pleistocene age. An effort should be made to study the cliffs along the southern side of the main volcano, for it seems likely, according to Allan Cox, that these will reveal the oldest of the visible lavas of the island.

Darwin was much impressed by what he saw of the eastern, younger part of Chatham Island, particularly by the profusion of small basaltic craters and the cones of palagonite tuff. Our own observations were confined to the extremely youthful flows of pahoehoe lava bordering the beautiful Sappho Cove and the area stretching northward to Cerro Brujo. The flows here are marked by many pressure mounds and ridges, and they consist of porphyritic olivine basalt devoid of feldspar phenocrysts. Cerro Brujo, a conspicuous headland on the coast, seems from a distance to be a complex cone built largely of palagonite tuff; it merits closer study for it appears to include steeply dipping and even overturned beds.

We cannot do better than quote from Darwin's account of the general appearance of the rest of the eastern part of the island.

. . . a bare, undulating tract, remarkable from the number, proximity, and form of the small basaltic craters with which it is studded. They consist, either of a mere conical pile, or, but less commonly, of black and red, glossy scoriae, partially cemented together. They vary in diameter from thirty to one hundred and fifty yards, and rise about fifty to one hundred feet above the level of the surrounding plain. From one small eminence, I counted sixty of these craters, all of which were within a third of a mile from each other, and many were much closer. . . . Small streams of black, basaltic lava, containing olivine and much glassy felspar, have flowed from many, but not from all of these craters.

He also noted many small, cylindrical collapse pits, particularly near the bases of the small craters.

It was in this eastern part of Chatham Island that Darwin first discovered what are now called cones of palagonite tuff, and their mode of origin intrigued him even more when he found other examples on James and Albemarle islands. His description, because of its historic interest, deserves to be quoted in full.

Towards the eastern end of the island, there occur two craters composed of two kinds of tuff; one kind being friable, like slightly consolidated ashes; and the other compact. This latter substance [palagonite] . . . is of a yellowish brown colour, translucent, and with a lustre somewhat resembling resin. . . . In a hand specimen, this substance would certainly be mistaken for a pale and peculiar variety of pitchstone; but when seen in mass its stratification, and the numerous layers of fragments of basalt, both angular and rounded, at once render its subaqueous origin evident. . . . This resin-like substance results from a chemical change on small particles of pale and dark coloured scoriaceous rocks. . . . The position near the coast of all the craters composed of this kind of tuff or peperino, and their breached condition, renders it probable that they were all formed when standing immersed in the sea. . . . I think it highly probable that much the greater part of the tuff has originated from the trituration of fragments of the grey, basaltic lavas in the mouths of craters standing in the sea.

Darwin also noted that in one of the two cones of well-bedded palagonite tuff the layers are cut by vertical dikes of "tuff like that of the surrounding strata, but more compact and with a smoother fracture." Perhaps these clastic dikes were formed by injection of water-soaked materials from below during seismic shocks.

One of the most celebrated features in the entire archipelago is Kicker Rock, known also as El Leon Dormiente, which stands 3 miles offshore, at the mouth of Stephens Bay. The precipitous cliffs of the main "rock" rise to a flattish top 486 feet above the sea; separated from this "rock" by a narrow, almost vertical-sided cleft, is a smaller, lower crag (Pl. 3). The adjacent sea floor falls abruptly to depths of 50 to 60 fathoms.

Darwin correctly described Kicker Rock as the remnant of a tuff cone, but he was mistaken in supposing that it marks the filling of a crater. It is composed of thinly bedded, buff-colored palagonitic ejecta, mostly of the size of sand and gravel. The bedding within the smaller crag dips southward at low angles; within the main part of the larger crag it lies horizontally, and at the northern end it dips northward at increasingly greater angles. These attitudes indicate that the crags are carved in a broadly arched pile of tuff, part of the flank of a very large cone. The flattish top of the main crag was not formed by marine planation; it merely reflects the bedding of the underlying tuff.

Petrography

Chesterman (1963) described a basalt from Sappho Cove as containing between 10 and 15 percent of olivine, accompanied by pale purplish augite. Presumably this is an alkali-olivine basalt. The specimen we selected for analysis (Table 2c, No. 103) came from near the western end of the island, between Progreso and Wreck Bay, and is probably representative of the lavas that form the main volcano. It also is an alkali-olivine basalt. It contains about 10 percent euhedral microphenocrysts of olivine ($2V = \sim 90°$), about 50 percent laths of medium labradorite up to 2 mm long, 30 percent ophitic, purplish-brown titanaugite, 5 percent ilmenite and magnetite, and 5 percent anorthoclase and opaque glass. It is essentially identical, texturally and mineralogically, to the dominant type of basalt on Hood Island, illustrated in Figure 3c.

A specimen of palagonite tuff from Kicker Rock (No. 114) was studied for us by Dr. Azuma Iijima. The palagonite itself, which makes up about two-thirds of the specimen, has a pale-yellow to brownish-yellow color and is almost isotropic. Its refractive index is approximately 1.600. The cement between the palagonite fragments consists of phillipsite (5 percent), chabazite (5 percent), calcite (5 percent), and analcite (15 percent). The silica-poor analcite occurs in very minute, isotropic crystals having an Si/Al ratio of 1.70, according to Saha's method (1959).

INDEFATIGABLE (SANTA CRUZ) ISLAND

Indefatigable Island, which is roughly oval in plan, measuring 20 by 25 miles across, is primarily a shield of basaltic lavas dotted,

KICKER ROCK

The remains of a tuff cone off the northern coast of Chatham Island. (Photograph by Charles Rick.)

McBIRNEY AND WILLIAMS, PLATE 3
Geological Society of America Memoir 118

particularly in its upper part, with many parasitic cones. Despite its large diameter, the summit peaks rise to elevations of only 2553 and 2385 feet; hence the slopes of the principal volcano are very gentle. No topographic maps are available for the interior of the island, and we were unable to examine aerial photographs. We are therefore all the more grateful to David Cavagnaro for his sketch map, reproduced here as Figure 5.

The profile of the island is markedly different from the much steeper profiles of the shield volcanoes of Albemarle and Narborough islands; it resembles more closely the profile of the main volcano of Charles Island. Were it not for the many large scoria cones on the summit plateau and on the flanks, the Indefatigable volcano would have the shape of an overturned saucer. The main volcano, like that of Charles Island, is probably in the "Mauna Kea stage" of development, but if there was ever a large summit caldera, developed in the earlier "Mauna Loa stage," it is now completely concealed. Many youthful scoria cones with well-preserved craters are present in the highlands, but these, instead of showing arcuate alignments, are grouped in an approximately east-west belt, roughly parallel to the recent fault scarps bordering Academy Bay and the faults that traverse Barrington Island (Pl. 4).

The few lavas we collected from the Indefatigable volcano are alkali-olivine basalts, mostly of pahoehoe type. Youthful flows, with normal magnetization, are well and widely exposed near Academy Bay and along the coast to the east; none of these contains large phenocrysts of plagioclase, but all carry phenocrysts of olivine and some are sufficiently rich in that mineral to be classed as picrites. A thick cover of soil and vegetation masks all but a few exposures in the interior of the island, but on the slopes of a parasitic cone near Los Pozos, north of Academy Bay, at an elevation of 1550 feet, we collected an olivine basalt with abundant large phenocrysts of plagioclase.

Red Hill, which lies a short distance inland from Conway Bay, at the northwestern end of Indefatigable Island, consists, at least in part, of palagonite tuff, a specimen of which was collected for us by Professor Charles Rick, and Eden Island, a short distance offshore, is also an eroded cone of thinly bedded palagonite tuff.

Two conspicuous, recent fault scarps cut the lavas around Academy Bay, and these, as previously noted, trend slightly south of east, in line with equally youthful fault scarps on Barrington Island. Several parallel fissures, presumably earthquake cracks, are said to be present in the country crossed by the trail leading from Academy Bay to the upland farms. Other faults and fissures, trending approximately east-west, cut the uplifted submarine lavas along the northeastern coast of the island.

YOUTHFUL SCORIA CONES IN THE WESTERN PART OF THE HIGHLANDS OF INDEFATIGABLE ISLAND

(Photograph by David Cavagnaro.)

Finally, note should be made of the fact that many lumps of pumice, some of them of fist size, have been washed up on the beach near the Angemeyer home at Academy Bay; further reference is made of these on page 113.

Petrography

We examined only a few specimens from Indefatigable Island, and most of these came either from near Academy Bay or from the northeastern coast. The former are probably representative of most of the lavas erupted subaerially to form the main volcano of the island; most of the latter are older, uplifted submarine lavas and have been described already (pages 16-17).

All of the lavas around Academy Bay are alkali-olivine basalts with as much as 5 percent normative nepheline. Some of them are sufficiently charged with olivine to be called picrites. A typical specimen selected for analysis (Table 2c, No. 1) comes from an outcrop between the laboratory building and the pier at the Darwin Station; it is illustrated in Figure 6b. It consists of slender laths of medium labradorite up to 1 mm long (~50 percent), olivine crystals up to 3 mm across, with an optic angle close to 90° (25 percent), subophitic reddish- and purplish-brown augite (10 percent), magnetite, ilmenite, opaque brown glass, accessory apatite, and interstitial zoned anorthoclase. Some olivine crystals show shadowy extinction, as if deformed, and nearly all contain inclusions of picotite. One specimen of basalt contains a single corroded crystal of calcic plagioclase, about 1.5 mm across; this may be a xenocryst, for it contains many inclusions of clinopyroxene and has a clear rim of sodic labradorite.

Fresh specimens are hard to find on the upper slopes of the island. Among those we collected, however, is an unusual basalt from the parasitic cone near Los Pozos, to which reference has already been made. This rock is illustrated in Figure 6a. It has many phenocrysts of sodic bytownite, about 0.5 cm long, with weak, normal zoning and sharp albite and pericline twinning. Euhedral phenocrysts and glomerocrysts of olivine (2V = ~90°) are less numerous; they exhibit shadowy extinction and most of them have rims of iddingsite and inclusions of picotite. The fine-grained intergranular groundmass consists of medium labradorite, clinopyroxene, and magnetite. Curious clusters of dark-brown spinel, about 0.5 mm across, are scattered throughout. Some of these spinels have rounded outlines; others are rectangular and appear to have inherited the forms of pre-existing minerals. The intergranular spaces within the clusters are occupied by bytownite, clinopyroxene, and other minerals of the groundmass.

DAPHNE ISLANDS

Daphne and Little Daphne islands lie north of Indefatigable Island and west of Baltra Island. Both are remnants of fairly young pyroclastic cones. Little Daphne is definitely the older of the two; indeed, erosion has reduced the original cone to an almost cylindrical stump encircled by precipitous cliffs. Even so, a shallow summit depression, mantled with *Scalesia* trees, remains to mark the crater.

The larger Daphne Island preserves much more of its initial conical form, most of the coastal cliffs being less than 50 feet high. It is composed of well-bedded, fine, buff and greenish-buff tuff, in part palagonitic, containing many angular lapilli and a few blocks of basalt. A large summit crater has a smaller one on its southern rim. Locally, palagonitization zones within the tuffs cut across the stratification, indicating that the alteration from the original sideromelane was at least in part postdepositional. Cavernous weathering is a striking feature of the lime-rich tuff.

Petrography

The only rock we examined microscopically is a lithic fragment from the tuff close to the rim of the main Daphne crater. This is an alkali basalt consisting of microphenocrysts of euhedral olivine in an intergranular matrix of sodic labradorite, augite, olivine, ilmenite, and magnetite. The olivine crystals, which reach 3 mm across, show no signs of reaction with the groundmass, and they carry numerous inclusions of picotite. Interstitial patches of anorthoclase are present in the groundmass.

DUNCAN (PINZÓN) ISLAND

Duncan Island measures only about 3.5 by 2.5 miles; nevertheless, in common with the still smaller, neighboring island of Jervis, it shows a greater variety of igneous rocks than do any of the larger islands of the archipelago. The summit of the island rises to an elevation of 1502 feet, and the original cone may never have risen much higher. It is apparent, however, that most of the western and southwestern parts of the volcano have been lost, partly by marine erosion and partly by downfaulting.

Chubb mistakenly called Duncan and Jervis islands "tuff volcanoes" devoid of lava fields. Both islands, however, and particularly Duncan, consist almost wholly of lava flows. Most of those on Duncan Island radiate from a center a short distance southwest of the summit, from a crudely circular depression that may mark the site of the principal crater. The slopes that descend to the coast appear to be constructional surfaces scarcely modified by erosion. They are thickly strewn with large boulders derived from the crusts of the latest flows; in fact, the boulders are so numerous and closely spaced that our helicopter landed only with great difficulty, and we ourselves had trouble in finding enough suitable space in which to camp.

The most conspicuous feature of the island is a steep-walled caldera northwest of the summit (Fig. 11). It measures approximately

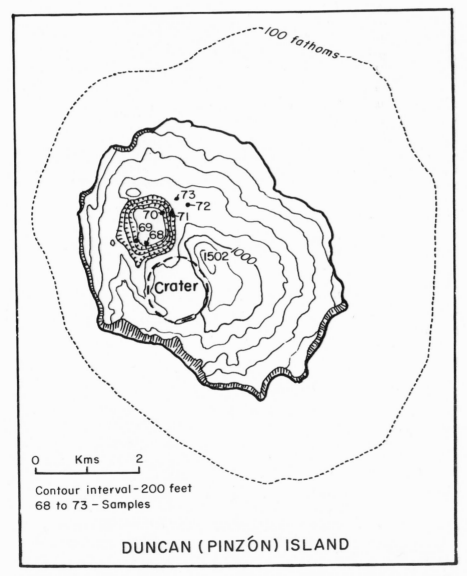

Figure 11. Map of Duncan (Pinzón) Island. (Topographic base *after* U.S. Hydrographic Office chart No. 5936, 1st ed., 1947.)

two-thirds of a mile across and between 300 and 400 feet deep. If we are correct in thinking that the crudely circular depression southwest of the summit of the island marks the main vent, then the caldera lies on the northwestern flank of the original cone, and the line connecting it with the principal vent runs parallel to one of the two dominant fracture systems of the archipelago. The absence of a rim of fragmental debris implies that the caldera was formed by engulfment.

The upper part of the steep walls on the northern, eastern, and southwestern sides of the caldera is formed by a single basaltic flow between 200 and 300 feet thick. On the southern wall and the adjacent bench there are thinner flows separated by reddish layers that are either fragmental beds or, more likely, the scoriaceous tops and bottoms of the flows themselves. The floor of the huge pit is littered with blocky basalt, almost completely buried by a thick mantle of inwashed silt and clay and wind-blown dust. Among the coarser debris in the alluvial fan that spreads onto the floor from the canyon to the south we found a few fragments of pumiceous trachyte, highly altered platy lavas, and opal.

The cliffs along the northwestern coast of the island were seen only from aboard ship. They reach heights of more than 300 feet and consist of abundant flows, much thinner than those on the walls of the caldera, with scoriaceous interbeds, and locally they are cut by a few thin dikes.

Two lavas from the upper part of the northeastern wall of the caldera were found by Allan Cox to have reversed magnetization.

Petrography

The lavas of Duncan Island exhibit a wide range in composition and deserve much more attention than we were able to devote to the small collection made during our brief reconnaissance. Excellent exposures in the caldera walls and in the coastal cliffs should be examined in detail, and the area bordering the southern side of the caldera should also be examined as the probable source of some of the unusual rocks we saw in the canyon entering the caldera from that direction.

The dominant rocks in the caldera walls, at least in their upper part, are strongly porphyritic, pigeonite-bearing icelandites (Fig. 12a). About half of their volume consists of phenocrysts of medium labradorite, many of which are zoned to andesine or oligoclase. Rare inclusions of olivine are present within a few of these phenocrysts, but they seem to be lacking in the remainder of the lava. Phenocrysts of augite reach a maximum length of 2 mm; some are zoned, showing a difference in extinction angle of as much as 6° between the core and the rim, but there is little if any corresponding difference in the

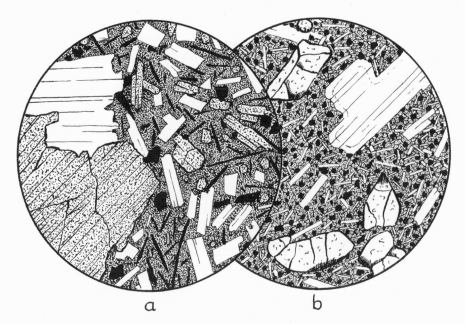

Figure 12. Microdrawings of Duncan Island lavas. Diameter of each field is
2 mm. (a.) Icelandite from eastern wall of the caldera (No. 70). Phenocrysts of
labradorite, with more sodic rims, and of augite. Groundmass is composed of laths
of andesine, granules of augite, pigeonite, and magnetite, plates of ilmenite,
and dark, "dust-filled," devitrified glass. (b.) Olivine basalt from the lower part
of the southern wall of the caldera (No. 68b). Phenocryst of labradorite with a
rim of andesine; also phenocrysts of partly oxidized olivine. Intergranular groundmass of labradorite laths with abundant granules of magnetite and clinopyroxene.

optic angle or refractive index. In these respects, the augites resemble
the large one from Barrington Island that we examined in greater
detail (p. 160). The hyalo-ophitic groundmass contains andesine and
two pyroxenes, a subcalcic augite ($2V_z = \sim45°$), and subordinate
pigeonite ($2V_z = \sim15°$). Long plates of ilmenite and granules of
magnetite are abundant. The glass is partly devitrified and contains
plumose microlites of clinopyroxene. Analyses of two specimens are
presented in Table 11 (Nos. 70, 71).

Porphyritic basalts are also present on the caldera walls, especially near the bottom. They closely resemble the icelandites in hand
specimens and hence are difficult to distinguish in the field. Under
the microscope, however, they are readily recognizable by their small
olivine phenocrysts and holocrystalline groundmass. Phenocrysts and
glomerocrysts up to 8 mm long make up about one-fifth of the volume.
Most of them consist of medium labradorite with rims of calcic
andesine; olivine follows in order of abundance; augite is least
plentiful. The plagioclase exhibits a wide range of size and form.
Some crystals have sharp, broken edges whereas others are rounded
and embayed; some show normal zoning, others show reverse zoning,

still others are totally unzoned. The intergranular groundmass is made up of sodic labradorite, pale-green clinopyroxene, oxidized olivine, and hematite. An analysis of a typical basalt (Fig. 12b) from the caldera wall is given in Table 2c (No. 68b).

Among the loose fragments strewn over the floor of the caldera we found one of pumiceous trachyte consisting of rare phenocrysts of oligoclase, green augite, and magnetite in a matrix of partly devitrified glass. Abundant vesicles are largely or wholly filled with chalcedony, and it may be that the trachyte has been hydrothermally altered.

JERVIS (RÁBIDA) ISLAND

Jervis Island lies close to the center of the Galápagos archipelago, and, although it measures no more than 1.5 miles, it contains a more diversified group of rocks than any other island, the lavas ranging from tholeiitic ferrobasalts through icelandites to siliceous hedenbergite trachytes.

The island is essentially a cluster of steep-sided, coalescing domes and stumpy flows built on an uneven floor of basaltic lavas and scoria (Fig. 13). One of the irregular depressions in the summit area may be a crater, but the others are simply low places between domical piles of lava. Our observations were directed mainly toward the younger, more siliceous lavas and the cones of fragmental ejecta along the northern coast; the underlying basaltic rocks, well-exposed in cliffs along the opposite coast, were examined only in a very cursory fashion. The blocky lavas that make up most of the island must have been discharged in an extremely viscous condition, forming steep-sided piles and thick, bulbous flows, some of which congealed on slopes of more than 45°. The highest point on the island (1203 feet) is the top of a blocky dome of siliceous lava from which thick flows spread northwestward to the coast (No. 68). Some lavas in the summit area contain abundant phenocrysts of plagioclase (for example, No. 45) ; others are darker, more basic flows containing fewer phenocrysts of plagioclase but many of augite (No. 44).

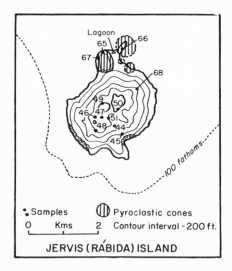

Figure 13. Map of Jervis (Rábida) Island. (Topographic base *after* U.S. Hydrographic Office chart No. 5936, 1st ed., 1947.)

Several eroded cones of fragmental ejecta lie at the northern base of this cluster of domes and flows. One of these cones forms a small, narrow peninsula at the northern tip of the island. It consists of reddish, basaltic agglutinate apparently erupted from a linear source a short distance to the south. At least one flow of scoriaceous basalt (No. 66) is intercalated between the well-bedded ejecta. Two smaller cones of agglutinate, partly buried by lavas originating near the summit of the island, are exposed along the coast between 100 and 300 yards south of the peninsula.

A much larger and younger fragmental cone lies to the west of the peninsula and the adjoining lagoon. This consists of well-bedded, gray and brownish basaltic scoria crowded with angular, lithic blocks of basalt but including also a few coarse-grained blocks of gabbro, eucrite, and more siliceous plutonic rocks, some of which exceed a foot across. Blocks of dunite and peridotite are notably absent.

Large crystals of augite were collected from the beaches of Jervis Island by Ochsner, during the California Academy of Sciences Expedition of 1906, but during our brief visit we saw none. Presumably the crystals came from the older basaltic lavas and scoria that underlie the siliceous flows.

One lava on the northwestern coast was found by Allan Cox to show normal magnetization; two others showed reversed magnetization.

Petrography

Lavas. The lavas of Jervis Island, as noted already, display the widest and most complete range of composition of any in the archipelago. Analyses of many samples are presented in Table 11. The oldest lavas that we sampled are ferrobasalts. A typical one (No. 44) from the southern slopes of the island, a short distance above the coastal cliffs, consists of phenocrysts of calcic labradorite with thin rims of andesine, subordinate augite, and coarse magnetite in a groundmass containing laths of calcic andesine. Intergranular olivine, hypersthene, augite, magnetite, ilmenite, interstitial anorthoclase, and accessory apatite complete the groundmass assemblage. Analcite and an unidentified zeolite (length-slow and straight extinction) are present in some of the vesicles. Another specimen (No. 45), from the overlying flow, differs in having more labradorite phenocrysts and late-crystallizing alkali feldspar, and its groundmass carries a little more olivine and augite.

Somewhat similar lavas are exposed higher on the slopes of the island. An eroded flow on the eastern side of the breached crater (No. 51) contains widely scattered phenocrysts of augite up to 15 mm in length. When viewed in thin section, the rock is seen to contain fragments of tightly intergrown calcic bytownite, some with labradorite rims, olivine, very coarse magnetite, and brownish prisms of apatite. Some plagioclase crystals are strongly corroded and rich in inclusions, and the large olivine crystals are embayed and surrounded by thin reaction rims. These fragmental crystals appear to have been derived from plutonic rocks such as those described below. The pilotaxitic groundmass of the enclosing lava consists of laths of andesine, very abundant magnetite, olivine, minute grains of clinopyroxene, patches of a zeolite, possibly natrolite, and interstitial anorthoclase.

The blocky lavas making up the central domes of the island (Nos. 49, 50) are more siliceous, some of them resembling rocks of intermediate composition from Iceland, recently described by Carmichael (1964) under the name "icelandites." About one-third of their volume is made up of phenocrysts and glomerocrysts of calcic andesine up to 6 mm long. Some phenocrysts are sharply euhedral and unzoned; others, of more calcic composition, are commonly broken and normally zoned to thin rims of sodic andesine. Clinopyroxenes are less abundant than plagioclase and form smaller phenocrysts. They include both hedenbergite ($2V_z = 49°$; $N_Y = 1.695 \pm 0.003$) and nearly uniaxial pigeonite, the latter forming slender, multiply twinned prisms that contrast with the stumpy crystals of green hedenbergite. Magnetite grains reach 0.5 mm across and account for about 5 percent of the total volume. The hyalopilitic groundmass contains randomly oriented microlites of clinopyroxene and oligoclase separated by colorless to pale-brown glass. In one specimen (No. 49), the groundmass is divided into two portions, one consisting of dark globular masses in which microlites of pyroxene and plagioclase are set in a dusty brown glass and the other consisting of almost colorless glass with the same proportion of microlites. Thin shells of magnetite commonly mark the boundaries separating the dark and light areas.

Hedenbergite trachyte lava occupies the highest part of the southern wall of the eroded crater and extends a short distance downslope to the south. A typical specimen (No. 48) is made up of sparse phenocrysts of andesine and oligoclase, a poikilitic green hastingsite ($Z\wedge c = 12°$; $N_Y = 1.690\pm000.3$), hedenbergite ($2V_z = 52°$; $N_Y = 1.718\pm0.002$), and magnetite. The fine-grained to glassy groundmass contains microphenocrysts of anorthoclase, clinopyroxene, magnetite, hematite, and abundant cristobalite in vesicles.

Rocks within the eroded crater are for the most part strongly altered, generally to opal. Feldspar phenocrysts are completely replaced by clay and the glassy groundmass by opaque, whitish opal. Clear opal and chalcedony line some of the vesicles, and many rocks are coated and veined with hydrothermally deposited silica minerals.

The lavas on the northeastern peninsula of the island are olivine-bearing basalts (No. 66). Phenocrysts amount to only about 5 percent of the volume, about half of them being plagioclase. Most of these plagioclase phenocrysts are sodic bytownite, but there are also corroded ones of more sodic composition that appear to have been derived from coarse-grained plutonic rocks such as those described below. Olivine is less abundant than augite and shows reaction relations to the groundmass, most of which consists of colorless to pale-brown glass containing abundant magnetite and a few patches of zeolite.

A slightly older lava from near the eastern end of the lagoon on the northern shore (No. 65) is similar to the basalt just described but contains less olivine. Many of the olivines are hollow and have thin reaction rims; some of the augite crystals have grown ophitically over the plagioclase, whereas others are more nearly euhedral and show hourglass zoning. The groundmass contains andesine, magnetite, and glass.

Plutonic Rocks. Coarse-grained blocks of varied composition were ejected from the cinder cone on the northwestern shore of Jervis Island. Their chemical composition parallels that of the lavas, and for every type of effusive rock there is a plutonic equivalent. The principal difference between the lavas and corresponding plutonic rocks, apart from grain size and texture, is the greater variety of mafic minerals and the relative scarcity of magnetite in the plutonic rocks. In the following notes, we describe a few representative specimens, but it should be emphasized that among the 15 rocks examined microscopically there is a continuous gradation.

Most common among the plutonic ejecta are olivine gabbros, and among those the most common consist of augite, hornblende, olivine, and sodic bytownite (Nos. 67-A, 67-B, 67-M, and 67-O). Anhedral grains of plagioclase are tightly intergrown and reach a maximum dimension of about 1 cm; they are normally zoned from sodic bytownite to rims of medium labradorite. Ophitic augite ($2V_z = 56°$; $N_Y = 1.687 \pm 0.002$) reaches 1.5 cm across. The hornblende is strongly pleochroic from reddish orange to pale yellow and has a negative axial angle of about 79°. It commonly forms euhedral overgrowths on augite. The olivine is intensely oxidized and in some rocks is completely replaced by opaque ores. Unoxidized cores have axial angles close to 90°. Magnetite is present as both a primary and a secondary mineral but is not abundant. Calcite, apatite, and sulfur are present in most samples in accessory amounts.

Hornblende shows the greatest variation in amount among the minerals of the olivine gabbros. Biotite is less common, but in one specimen (No. 67-C) it makes up about 5 percent of the volume, occurring as coarse, reddish-brown flakes up to 5 mm across

($2V_X = 5°$ to $10°$; $N_z = 1.636 \pm 0.002$), accompanied by olivine, augite, hornblende, and bytownite-labradorite. Little if any magnetite is present.

With increasing iron content and more sodic plagioclase, the foregoing plutonic rocks grade into olivine-bearing ferrogabbros (Nos. 67-I, and 67-J), consisting of sodic labradorite or calcic andesine, green salite ($2V_z = 54°$; $N = 1.699 \pm 0.002$), hornblende, oxidized iron-rich olivine (Fa about 50), coarse apatite (2 to 3 percent), brown biotite, secondary magnetite, calcite, and reddish-brown glass. The hornblende is strongly pleochroic from dark brown to brownish green and commonly has cores of clinopyroxene. Anorthoclase is present in subordinate amounts between the larger grains of plagioclase.

The most siliceous plutonic rocks that we collected are single specimens, one of hortonolite leucodiorite (No. 67-D), which is illustrated in Figure 14c, and one of quartz-bearing syenite (No. 67-N). These rocks differ mainly in the proportions of their dark and light constituents. The leucodiorite is made up mostly of calcic oligoclase, which, to judge from the abundance of pericline twins and the refractive index ($N_Y = 1.540$), is probably rich in potassium. This strongly twinned plagioclase is rimmed with untwinned anorthoclase, which also occurs as small, discrete crystals. Hedenbergite ($2V_z = 52°$; $N_Y = 1.713 \pm 0.002$) is the most plentiful mafic mineral. It is accompanied by ferrohortonolite (Fa_{86}) and ferrohypersthene (Fs_{73}) but by little if any hornblende or biotite. Apatite is an abundant accessory.

The quartz-bearing syenite contains more oligoclase, the potassic nature and optical properties of which are shown in Table 15. The hedenbergite resembles that in specimen 67-D; its composition is also given in Table 16. Two other pyroxenes are present but in very subordinate amounts. Ferropigeonite ($2V_z = \sim 13°$; OP normal to 010) forms the cores of some hedenbergite crystals; ferrohypersthene (Fs_{72}) forms small, discrete grains. Quartz constitutes between 1 and 2 percent of the total volume. Accessory minerals include sphene, apatite, and magnetite, in that order of abundance.

Analyses of the bulk compositions of representative specimens of the main varieties of plutonic ejecta are presented in Tables 13 and 14, and their petrological significance is discussed on pages 148-152. Five of the rocks are illustrated in Figure 14.

JAMES (SAN SALVADOR) ISLAND

James Island is one of the most fascinating in the archipelago; nevertheless, every visiting scientist since Darwin has focused his attention on the attractive and readily accessible geologic and biologic

Figure 14. Microdrawings of plutonic ejecta from Jervis Island. Diameter of each field is 6 mm. (a.) Olivine gabbro (67-B). Olivine mainly replaced by magnetite. The augite plate near the center has a reaction rim of reddish-brown hornblende. Laths of bytownite with rims of labradorite. (b.) Olivine eucrite (67-O). Olivine almost wholly replaced by magnetite. All of the augite is in optical continuity and is part of a plate 1 cm across. Large laths consist of bytownite. (c.) Ferrohortonolite leucodiorite (67-D). Calcic oligoclase, partly rimmed with anorthoclase; hedenbergite; ferrohortonolite partly replaced by magnetite; accessory apatite. (d.) Biotite gabbro with olivine rimmed by augite and hornblende and large interlocking anhedral grains of bytownite (67-C). (e.) Quartz syenite consisting of potassic oligoclase, ferroaugite, ferrohypersthene, fayalite, minor quartz, sphene, and apatite (67-N).

features around James and Sullivan bays. The interior of the island remains virtually unexplored.

The island is 21 miles long in a northwest-southwest direction and about 15 miles across. It consists primarily of a major volcano that occupies most of the northwestern part, rising to a height of 2974 feet, and a swarm of younger, relatively minor cones, mostly built along northwest-trending fissures, from which copious floods of basaltic lava have been erupted during Holocene time. Probably none of the visible lavas on the island date back as far as Pliocene time, and there seem to be no uplifted submarine flows comparable with those on Hood, Barrington, Indefatigable, and Baltra islands.

Many eruptions have been reported during the last century from a conspicuous cone on the southwestern side of the island (Chubb, 1933, p. 16). A flank eruption was seen in 1897 on the southeastern side, and Slevin (1959, p. 10) says that when he visited the island in 1906 fumaroles were active near the summit. No one who lands on the island or flies over it can fail to be impressed by the vast sheets of barren, black basalt, particularly around James and Sullivan bays, and all along the southern slopes of the island. So youthful are these lavas that there are scarcely any high cliffs along the coast, and nowhere in the interior are there signs of even a moderate degree of erosion. Everywhere the topography consists of essentially unmodified constructional forms; in countless places on the lower slopes, delicate surface textures of the lavas and other fragile features, such as miniature spatter and driblet conelets, stand as if formed yesterday.

The main volcano that makes up most of the northwestern part of the island has the crude form of an overturned boat, measuring 9 by 7 miles (Pl. 5). It reminds one of the well-known volcano of Hekla in Iceland, and like that volcano it was built by repeated eruptions, chiefly of basaltic lava with minor amounts of fragmental debris, from a series of vents aligned along parallel, closely spaced fissures. Minor topographic details are obscured by vegetation, but there is no reason to suspect the presence of a large summit crater, much less a summit caldera. The lavas on the southern slopes of the volcano are in general less vegetated than those on the opposite slopes, but even where vegetation is dense the pahoehoe crusts of the flows are well-preserved.

After the main volcano had been built, activity shifted to the lower flanks of the island, especially to the southern and eastern sides, where extensive floods of barren lava are dotted with innumerable small cones of tuff, scoria, and spatter. In describing the products of these later eruptions, we direct attention first to the area around James Bay, an area to which Darwin himself paid particular attention. Cape Cowan, which borders the northern end of the bay, is a cone of yellowish and brownish sideromelane and palagonite tuff heavily charged with angular, lithic blocks and lapilli of varitextured basalts. It measures a mile across the base and rises to a height of 932 feet. The summit crater, which is breached on the coastal side, is fairly well preserved.

Albany Island, half a mile offshore, is the arcuate remnant of a small lava-scoria cone. Half a mile to the north, the remnants of a deeply eroded cluster of coalescing lava-scoria cones form a narrow,

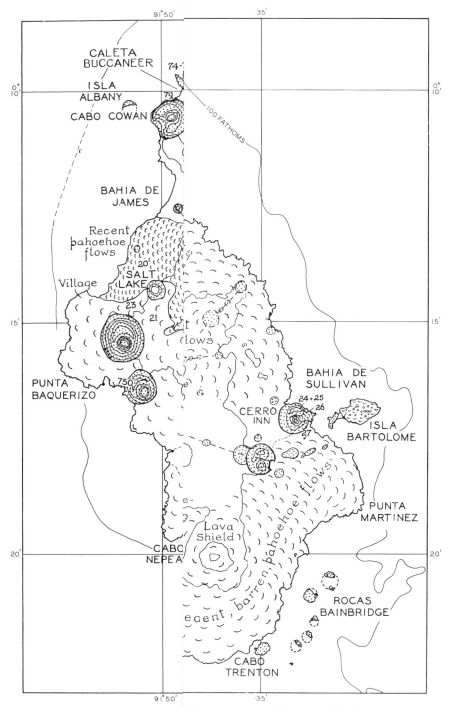

Stippled areas in with vegetation. (Topographic base areas denote Holocene 6, 1st ed., 1947.)

cliffed peninsula. These eroded cones are grouped along a fissure directly in line with the long axis of the major volcano of James Island (Pl. 5). They are undoubtedly the cones illustrated and vividly described by Darwin, and it must have been here that he gathered several of the unusual rocks long afterward studied by Constance Richardson (1933). It should be noted in this connection that Darwin referred to the locality as Fresh-water Bay; nowadays it is known as Buccaneer Cove. During our brief visit we noted many gabbroic xenoliths within both the beds of scoria and the intercalated flows. The lavas on the adjacent slopes of the major volcano contain only sparse phenocrysts of plagioclase, but some of those at higher elevations are extremely rich in phenocrysts of olivine.

The southern part of James Bay is bordered by spectacular sheets of barren pahoehoe lava that cover approximately 8 square miles, descending to the coast along a front of 2.5 miles. These flows, although slightly cliffed by marine erosion, must be very young, for even their most delicate surface features are still preserved (Pl. 6). Pressure ridges and collapse depressions are abundant, and some narrow cracks in the lava crusts are coated with small, ovoid balls of spatter. The source of these youthful flows lies about 2.5 miles inland, at an elevation of approximately 1000 feet, on the southern flank of the major volcano.

Immediately to the south of this field of barren lava lies Salt Lake Crater, described by Darwin in his journals. Here one sees a small, saline lake on the floor of a steep-walled crater between 400 and 600 yards across at the rim. The upper part of the walls and the surrounding cone consist of brownish, partly palagonitized lapilli tuffs containing angular blocks of basalt, some as much as 4 feet in diameter. The lower and steeper parts of the walls, in contrast, consist of many thin flow units of pahoehoe lava, no doubt equivalents of the tuff-covered lavas that surround the cone on all but its northern side. We think that the tuff was formed by phreatomagmatic explosions through the lavas and that after the explosions ended the magma retreated, bringing about an engulfment that deepened and widened the crater. The present surface of the salt lake is said to be about 30 feet below sea level; hence we suppose that the lake is fed principally from the coast through tubes and fractures in the pahoehoe lavas. Locally, according to Hector Egas, owner of the property, fresh water seeps from the walls of the crater to augment the marine supply.

A thick sheet of well-bedded, brownish tuff, much of it altered to palagonite, mantles the pahoehoe lavas south of Salt Lake Crater. Some of these ejecta were blown from the crater itself, but most of them probably came from the much larger cone a short distance farther south. This cone, which reaches a height of 1295 feet and measures more than 1 mile across the base, is made up of weakly indurated palagonite tuff and dark basaltic scoria that carry abundant angular blocks of varitextured basalt. Locally, irregular zones of palagonitization cut across the regular stratification of the ejecta. A similar, but smaller, cone rises a short distance to the south-southeast, presumably on a common fissure (Pl. 5).

All of the southern coast of James Island and the region stretching inland for at least 3 miles is a barren waste of youthful lavas dotted with perfectly preserved spatter and driblet cones. Some lavas are of aa type, but most are pahoehoe flows marked by many pressure ridges and collapse depressions. All were discharged from vents aligned along northwest-trending fissures on the southern flank of the island's main volcano.

Near the southeastern end of the island, about 2 miles inland from Cape Trenton, there is a broad, shieldlike massif of basaltic lava, probably the principal vent from which were erupted the fresh pahoehoe lavas that cover about 20 square miles, spreading at least 3 miles westward from Cape Trenton and northward to Sullivan Bay. The glistening, tarlike crusts of these flows are virtually devoid of vegetation; in places, they are smooth and flat or gently undulating; elsewhere they are humped into steep-sided pressure ridges and traversed by countless tubes. Their age is probably to be measured in hundreds or thousands rather than in tens of thousands of years.

Two large cones dominate the landscape around Sullivan Bay, one of them, Cerro Inn, close to the shore and the other about a mile inland, to the southwest. Cerro Inn consists of thinly bedded, buff-colored palagonite tuff containing angular fragments of basalt, a few of them 4 feet across. There are no large bombs and apparently no plutonic ejecta. A few thin, irregular dikes cut the tuffs in the coastal cliffs. The cone seems to have been built by eruptions from three coalescing craters rather than from the single one shown on topographic maps, and the eruptions were all of phreatomagmatic type, having resulted from explosions of basaltic magma that rose through water-saturated rocks. The cone a mile or so to the southwest seemed from a distance to be of similar origin. It is cut by a conspicuous rift that trends west-northwest, approximately in line with at least two smaller cones, a lava shield, and an unusually large collapse depression (Pl. 5).

Soon after explosive activity ended at Cerro Inn, a few small scoria cones developed close to its southern base. Then followed the discharge of pahoehoe lavas that flooded all of the southeastern end of James Island, to which reference has already been made. Still later, activity was resumed at Cerro Inn. Two steep-sided mounds of basaltic lava and agglutinate were built on an east-west fissure in the breached, eastern flank of the cone, as may be seen on Plate 7. Short flows of thin-shelled, brown-crusted pahoehoe lava poured through

ROPY SURFACES OF HOLOCENE PAHOEHOE LAVAS CLOSE TO THE
SOUTHERN SHORE OF JAMES BAY, JAMES ISLAND

McBIRNEY AND WILLIAMS, PLATE 6
Geological Society of America Memoir 118

RECENT CONELETS OF LAVA AND AGGLUTINATE ON THE EASTERN FLANK OF THE PALAGONITE-TUFF CONE OF CERRO INN, BORDERING SULLIVAN BAY, JAMES ISLAND

the breach, some of them spreading to the coast, overriding the margins of the earlier, black-topped pahoehoe flows. Close to the northeastern margin of this younger lava field there are tunnels partly coated with opal, presumably deposited by siliceous springs.

A few miles north of Sullivan Bay there is an extensive area of youthful flows related to a group of cones aligned along fissures that trend northwestward, an uncommon orientation for eruptive fissures in the archipelago.

Bartholomew Island, according to Jacques Laruelle and his colleagues (1964), consists of four distinct parts. The eastern, and by far the largest, part, which rises to an elevation of 359 feet, is essentially a plateau built of youthful lava flows dotted with many small cones of spatter and driblets. At its western base there is a narrow strip of land covered mainly with young pahoehoe flows, and west of this, forming the lowest part of the island, is a vegetated strip of beach and dune sands composed of shelly and basaltic debris. The cliffs at the western end of the island and the picturesque offshore stack that rises from Sullivan Bay are composed mainly of dark-green and reddish-brown basaltic tuffs and breccias, in part altered to palagonite, interbedded with a few thin flows of basalt. These are probably the oldest rocks on the island, and they seem to be the remnants of a group of deeply denuded, coalescing cones. Two narrow, dikelike ridges of reddish, basaltic scoria and agglutinate rise above the "sea" of young pahoehoe lavas on the adjacent mainland; these may be other remnants of cones built along the same northeast-to north-northeast-trending fissures.

The small islands known as the Bainbridge Rocks, a short distance from the southeastern coast of James Island, and the island close to Cape Trenton are remnants of palagonite-tuff cones, the arcuate outlines of some of them marking the original crater rims.

It should be noted in conclusion that many waterworn lumps of pumice, some fist-sized, were found by Professor Wyatt Durham on the beaches bordering Sullivan Bay on the south. Unfortunately, we cannot say whether the pumice is of local, Galápagos origin or was drifted from the coasts of Central or South America or from the Revillagigedo Islands off the coast of Mexico.

Petrography

Olivine basalts and picrites, which are well-exposed along the northern coast, are probably the most abundant of the older lavas of James Island. An analysis of a typical picrite is given in Table 2c (No. 76). About one-third of the specimen is made up of phenocrysts of magnesian olivine, accompanied by a few broken and partly resorbed phenocrysts of calcic labradorite. The fine-grained groundmass consists of small laths of calcic labradorite, clinopyroxene, olivine, magnetite, and a little glass.

Augite phenocrysts are rare among the older lavas of the island. In one olivine-rich specimen, however, augite forms small oikocrysts in the groundmass, similar to those in a basalt from Barrington Island illustrated in Figure 3. This unusual texture was also observed in a block of basalt from the tuff cone of Cerro Inn, bordering Sullivan Bay. Analysis of the Cerro Inn basalt (Table 2c, No. 24) shows it to be strongly alkaline, with almost 6 percent normative nepheline. It differs from the analyzed picrite principally in its lower content of magnesia.

Some of the older lavas, by diminution in the amount of olivine, grade into iron-rich tholeiitic basalts. Although olivine is still not uncommon in these rocks, the most abundant phenocrysts are broken and corroded phenocrysts of labradorite, many of which show a more calcic zone between a core and a rim of slightly more sodic labradorite. Phenocrysts of augite are present in a few of these ferrobasalts, but their total volume is small. The groundmass is characterized by seriate andesine, along with augite, abundant magnetite, and acicular apatite (Fig. 15b). Chemical analyses of three specimens are presented in Table 10 (Nos. 78, 79, and 81). These lavas resemble some of those of intermediate composition from Jervis and Duncan islands; they are among the most iron-rich lavas of the archipelago. Each of the analyzed specimens contains a small amount of olivine, apparently in equilibrium with the groundmass, despite the fact that their normative composition shows a little quartz.

The youngest lavas of James Island have already been well described by Richardson (1933). Those we collected from the large flow at James Bay are very much like those bordering Sullivan Bay; all are intergranular alkali-olivine basalts containing deep-brown, ophitic augite, calcic labradorite, magnetite, ilmenite, and a little brownish glass. The principal variations are in the abundance of olivine phenocrysts and the coarseness of the groundmass. The analyzed sample (Table 2c, No. 20), from the recent flow at James Bay, is probably of the same type as that described by Lacroix and Richardson, even though the chemical composition cited by Lacroix differs from that determined by Aoki. We believe that this difference, which is especially marked in the alkali content, merely reflects the more accurate determination by the flame photometer method used by Dr. Aoki. The lava in question resembles very closely that from Darwin Station on Indefatigable Island, illustrated in Figure 6b.

Gabbroic fragments, which are common in the lavas and fragmental ejecta at Buccaneer Cove, are less varied in composition than are those on Jervis Island. Most of them contain large crystals of plagioclase, between calcic labradorite and sodic bytownite in composition, and magnesian olivine in various stages of oxidation, together with large oikocrysts of brownish augite. In many thin sections, all of the clinopyroxene has a uniform optical orientation (Fig. 15a) ; it probably grew at a late stage from the interstitial liquid between loosely packed accumulations of olivine and plagioclase crystals. Since many of the plagioclase crystals enclose round grains of olivine, it appears that the sequence of crystallization was first olivine, then plagioclase, and finally augite. Primary ore minerals are notably scarce, but in some strongly oxidized blocks the olivine is completely replaced by opaque ore and hematite, and the margins of the augite crystals are reddened.

One gabbroic specimen contains about 2 percent hypersthene, but hornblende and biotite, which are plentiful in some of the gabbros from Jervis Island, were not detected among the James Island rocks. Crystals of plagioclase, many of which reach 1 cm in length, commonly show a band of reversed zoning, similar to that seen in the plagioclase phenocrysts of some of the older lavas of the island. A chemical analysis of a typical cumulate eucrite from the pyroclastic cone bordering Buccaneer Cove is given in Table 5 (No. 74).

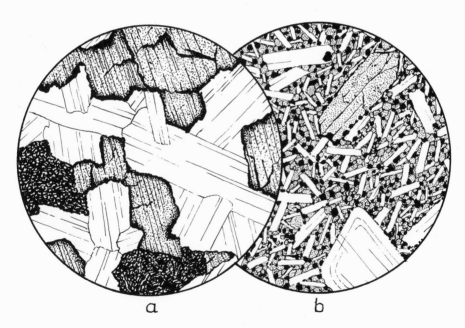

a   b

Figure 15. Microdrawings of James Island rocks. (a.) Olivine-augite gabbro (No. 74-F) from Buccaneer Cove. Diameter of field is 6 mm. The olivine is almost wholly replaced by magnetite and hematite; all of the augite is optically continuous; the feldspar laths range from calcic labradorite to sodic bytownite. (b.) Ferrobasalt from the northeastern slope of the island (No. 79). Diameter of field is 2 mm. Phenocrysts of labradorite and augite; intergranular groundmass of andesine laths and granules of augite and magnetite. (Analysis in Table 11).

Special effort was made to find the soda trachyte reportedly collected by Darwin from James Island and carefully described by Richardson (1933), but we found nothing remotely resembling this unusual rock. Dr. S. O. Agrell furnished us with a thin section of Darwin's specimen, now in the Sedgwick Museum, Cambridge, and this is almost identical with the specimen meticulously described by Richardson. It is, however, quite unlike any rock we found in the entire archipelago; hence, until the presence of trachyte on James Island is confirmed, we cannot exclude the possibility that the specimen described by Richardson was erroneously included in Darwin's collection.

A sample of palagonite tuff from Salt Lake Crater was examined for us by Dr. Azuma Iijima. Less than 5 percent of the tuff now consists of fresh glass, and this, which has a greenish to brownish-gray color and refractive index of 1.606 ± 0.002, encloses a few minute crystals of olivine and plagioclase. Roughly one-half of the tuff consists of palagonite. Most of this is clear and isotropic, with an index of about 1.700, and varies in color from golden yellow through orange to reddish brown. Along the margins of individual fragments and around the walls of vesicles, the palagonite becomes turbid and birefringent. The fillings of the vesicles and the cement between the fragments of tuff consist of the following minerals: phillipsite (8 percent); analcite, $N_x = 1.496$; $N_z = 1.498$); calcite (10 percent); and chabazite (7 percent $\omega = 1.494$; $\epsilon = 1.492$; the outer part of some zoned crystals has an index of about 1.480).

ALBEMARLE (ISABELA) ISLAND

Albemarle, which is by far the largest island in the archipelago, is the only one previously studied in detail. It is shaped somewhat like the letter J, measuring about 70 miles long in a north-northwest direction and 40 miles across the base. It consists of six huge, coalescing volcanoes, five of which are active shield volcanoes in the mature stage of development; the sixth, at the northern end of the island, has been largely destroyed by faulting and erosion. Each shield volcano has an enormous caldera at the top, around which are circumferential fissures, emphasized by rows of spatter and scoria cones. Countless flows of lava have poured down the flanks of the shields from these arcuate fissures, most of them to reach the sea. Other lines of spatter and scoria cones follow radial fissures on the flanks of the shields, and many of these cones are breached on their lower sides because of downslope extension of the rifts during eruptive activity. In addition, many small cones are aligned along roughly parallel lines that link the major volcanoes. Few of these parasitic cones exceed 500 feet in height and most are very much smaller. Their total number is probably more than 2500. Large cones of scoria, comparable with those on Charles and Indefatigable islands, are conspicuous by their absence; such cones seem to form only during the postmature stage of development of the major volcanoes.

The five great shield volcanoes of Albemarle, taken together, constitute one of the most active volcanic regions in the world. At least 14 eruptions have taken place here since 1911. Five of these have occurred on Cerro Azul since 1934, three on Sierra Negra since 1911, one on Alcedo in 1954, and five on Wolf since 1925. Only Darwin volcano has had no reported eruptions. We do not doubt, however, that on Darwin, as on each of the other shield volcanoes, many eruptions have taken place during historic times without a written record. Because of the great height and size of the volcanoes, and the scarcity of inhabitants in the vicinity, many unnoticed eruptions must have occurred within the deep summit calderas, and many flank eruptions, although possibly seen by chance from fishing boats, may never have been reported. All of the visible lavas of the five great shields were probably discharged during and since late Pleistocene times.

Chubb (1933) thought that three phases could be distinguished in the history of each of the shield volcanoes. During the first phase, he thought, activity was mainly explosive; then followed minor eruptions that built small parasitic cones; and lastly came a phase that still continues, characterized by copious eruptions of lava from fissures on the flanks. Wolf (1895) thought that initial submarine eruptions, typified by the accumulation of palagonite tuffs, were followed by regional uplift and that subsequent subaerial eruptions were marked especially by discharge of lava flows and the growth of many scoria cones. Banfield and his colleagues (1956), on the other hand, thought that "there was apparently an earlier stage mainly characterized by lava accumulation and building of the main craters, followed by a later stage in which pyroclastic material predominated. However, the greater part of the very fresh, lesser flows now visible at the surface are later than much if not most of the pyroclastic material." In our opinion, all five shield volcanoes on Albemarle Island, like the neighboring Narborough shield and the shields of Hawaii, were built almost wholly by closely spaced outpourings of fluid lava, at first mainly from summit vents but later from arcuate fissures around the summit calderas and from fissures on the flanks. Explosive activity contributed relatively little to visible parts of the growing shields at any stage; it served only to build chains of small spatter and scoria cones along the eruptive fissures. Had explosive activity predominated during early stages of growth, far more fragmental ejecta would be visible on the walls of the calderas.

Our observations on Albemarle Island were made during a day at Cape Berkeley, another on Beagle and Tagus cones, and during helicopter flights over Alcedo volcano. Fortunately, Banfield and his colleagues have already furnished a well-illustrated account of the five shield volcanoes; it must suffice, therefore, to summarize their descriptions and add a few observations of our own, beginning with an account of the southern volcanoes and proceeding northward to Cape Berkeley.

Cerro Azul Volcano

Although this is the smallest of the five shield volcanoes of Albemarle Island, it rises to a height of 5540 feet. The lower slopes are covered mainly by lavas whereas the higher slopes are covered mainly by pyroclastic ejecta. The summit caldera is slightly more than 2.5 miles long in an east-northeast direction and 1 mile or so across; its walls are unusually steep and have a complex, scalloped outline (Pl. 8). A bench on the walls of the caldera denotes repeated subsidence, as in other Albemarle calderas, and there are circumferential lines of spatter and scoria cones beyond the caldera rim. Also, as the aerial photograph (Pl. 8) shows, many radial fissures cut the outer slopes of the shield. Solfataric activity has continued for a long time within the caldera, and at least five lava flows have been discharged since 1943.

Sierra Negra Volcano (Volcan Grande)

This volcano, which reaches a height of 4890 feet, is markedly elongated in an east-northeast direction, in line with its neighbor, Cerro Azul. The caldera, which is elongated in the same direction, measures approximately 6 by 4 miles from rim to rim; nevertheless, it is only about 350 feet deep and hence is much the shallowest of all the Albemarle calderas. The outer slopes of the shield average less than 5°, diminishing to about 2° near the base. Above an elevation of about 500 feet, according to Banfield and his colleagues, lavas disappear and pyroclastic ejecta predominate. An approximately north-south, sinuous rift crosses the floor of the caldera near its western side, and close to it there is a large solfatara with conspicuous deposits of sulfur.

Recent lavas seem to be restricted to the outer slopes of the volcano. They issued from vents not far from the northern rim of the caldera where numerous small cones are aligned along circumferential fissures. Richards (1962) lists only three eruptions, one in 1911 or 1912, a second in 1948, and a third in 1953-1954. Many other eruptions must have passed without notice. The last outbreak took place

CERRO AZUL CALDERA, ALBEMARLE ISLAND

The scalloped margins of the caldera are surrounded by circumferential fissures along which numerous small scoria and spatter cones are aligned. Similar cones are abundant on the flanks of Cerro Azul volcano, some arranged along roughly radial fissures and others along a west-northwest-trending line. Countless recent flows have been erupted from both the straight and the arcuate fissures. The conspicuous bench on the caldera walls indicates that collapse took place in at least two stages. A large scoria cone rises from the fracture at the base of the eastern wall. (Photograph by U.S. Air Force, 1946.)

in 1963-1964 and was still in progress, although apparently drawing to a close, during our visit early in 1964. We are much indebted to Dr. David Snow for the following account of this eruption.

The eruption started at 10:30 p.m. on April 13, 1963, when a huge red mushroom cloud thousands of feet high rose into the air, forming a spectacular sight from Academy Bay, about 40 miles away. The cloud soon spread out laterally and became obscured by dark clouds gathering around it. It remained visible from Academy Bay for a few days. On June 27, I visited the south coast of Elizabeth Bay and found that two lava flows had reached the sea from an area of cones and craters on the north side of Sierra Negra at an estimated height of about 3,000 feet, and some 5 miles inland. The eastern flow had stopped on reaching the shore, but was steaming close to the sea; its surface was very similar to that of the 1963 flow from Volcan Wolf. The western flow had pushed out to sea for about 200 yards over a front of about a quarter of a mile. It was of hard, rope-like lava and was covered by a layer of white salt, with sulfur in places, the fine crystals being still intact. From a distance, the lava resembled a long, dazzling white beach. The eastern flow reached the coast not far, perhaps a mile or two, west of the two small islands in Elizabeth Bay, and the western flow some two miles east of Punta Moreno. The crater was still smoking slightly in early October 1963, but did not glow at night.

Fumes were still to be seen when our expedition sailed along the eastern coast of Albemarle Island in January 1964.

Alcedo Volcano (Volcan Calderón)

Although this volcano rises to an elevation of 3700 feet and has a caldera 4 miles wide, it is the smallest of the five shields of Albemarle Island, and it seems to have had fewer historic eruptions than any of the others. Most of its flanks, and indeed the summit caldera, are green with vegetation. There are, however, many extensive sheets of fresh lava on the flanks of the shield, particularly in and near the saddle separating Alcedo from Darwin volcano. Aa flows from flank fissures are also common around the base of the volcano, but flows from circumferential fissures around the caldera appear to be relatively scarce. Arcuate lines of spatter and scoria cones are in fact much less conspicuous than on any of the other four shield volcanoes.

A well-defined bench on the caldera wall testifies to at least two collapses. It stands about 500 feet above the innermost pit, on its northern and northeastern sides. The innermost pit is crossed by a narrow, gaping fissure that trends east-northeast, and on the southeastern wall of the caldera, approximately 300 feet above the floor, there is a vigorously boiling pool whose vapors smell strongly of hydrogen sulfide. Solfataric activity must once have been much more widespread in this part of the caldera, to judge by the extensive bleaching and discoloration of the lavas and the presence of scattered incrustations of sulfur.

The only recorded eruption of Alcedo volcano took place in November 1954, following shortly after an uplift of Urvina Bay at its northwestern base (*see* pages 101-102). Richards (1962) reports that a fissure opened on the north-northeastern flank of the volcano at an elevation of about 2000 feet. Lava may also have been erupted within the caldera, but if so we failed to detect it during our flights across the volcano.

Banfield and his colleagues thought that explosive activity lasted longer and was more widespread on Alcedo volcano than on any of the other major volcanoes, and we observed from the air that much of the outer slopes is mantled with pale-colored ejecta. Fortunately, Allan Cox was able to revisit the volcano in 1965. Around the southeastern base of the volcano, adjoining the low-lying Istmo Perry, which separates it from Sierra Negra, he observed a layer of pumice more than 10 feet thick, locally cut by gullies several yards across. The largest lump of pumice that he saw measured about 10 inches in length. Subsequent study has shown that the pumice is of trachytic composition, indicating that magmatic differentiation has proceeded much further at Alcedo volcano than at any of the other volcanoes of Albemarle Island. The top of the pumice layer has been slightly weathered so that the whitish color of the fresh ejecta has been changed to a reddish tint; nevertheless, the pumice mantles the slopes of the volcano and is only slightly eroded; it must have been erupted in very recent times. Perhaps no other volcano in the archipelago has discharged as much trachytic material.

Darwin Volcano (Formerly Mount Williams)

This almost perfectly symmetrical volcano measures 13 miles across the base and rises to a height of 4350 feet. Its slopes average about 18°, but near the top and bottom they are much gentler, as shown in Figure 16. The circular caldera is 3 miles wide and from 600 to 700 feet deep. An arcuate, flat-topped bench along the southern wall indicates that engulfment of the caldera took place in at least two stages. Two small collapse pits and two partly buried conelets are present on the bench (*see* Pl. 9). The floor of the caldera is covered by barren flows of lava; the vents of some of them can be noted close to the eastern and northeastern walls. On the broad, flattish bench surrounding the caldera rim, just as on Wolf volcano, there are many arcuate lines of spatter and scoria cones from which innumerable flows have descended the outer slopes. Many radial fissures on the flanks of the volcano have also produced copious

Figure 16. The five great calderas and shield volcanoes of Albemarle (Isabela) Island, together with some of the major rift zones and youngest lava flows. (*After* Banfield and others, 1956.)

flows, and one of these fissures extends beyond the base of the volcano to cut Tagus Cone, as described in the following section.

Tagus and Beagle Cones

Tagus Cove is one of the most sheltered and frequented anchorages in the archipelago. The *Beagle* anchored there in 1835, and the *St. George* in 1924, and both Darwin and Chubb spent considerable time exploring the adjacent area. Tagus Cone, as may be seen on Plate 10, has a complex form, but Beagle Cone and its diminutive neighbor are very simple structures. All three cones lie at the west-southwestern base of Darwin volcano, and the radial fissure from this volcano, which intersects Tagus Cone, would, if prolonged, pass through the center of the Narborough volcano. Similarly, if the line

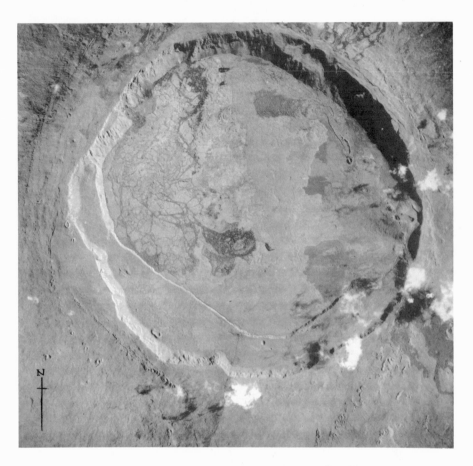

DARWIN CALDERA, ALBEMARLE ISLAND

Bench on the walls of the caldera and lines of scoria and spatter cones arranged along circumferential fissures around the caldera rim are shown. Two small collapse pits and the arcuate remnants of two scoria cones can be seen on the wide bench along the southern wall of the caldera; also a north-northwest-trending the eastern wall of the caldera. A recent flow, marked by a conspicuous lava gutter, issued from an intra-caldera cone on the northeastern part of the caldera floor. (Photograph by U.S. Air Force, 1946.)

BEAGLE (LOWER) AND TAGUS (UPPER) TUFF CONES,
WESTERN COAST OF ALBEMARLE ISLAND

Basal diameter of Beagle Cone is approximately 2 miles. The caldera of
Darwin volcano lies 5.5 miles to the northeast. Recent lavas from fissures on
the flanks of Darwin volcano have largely surrounded the tuff cones. One radial
fissure on Darwin volcano crosses the northern part of Tagus Cone, where it is
marked by a line of small lava-agglutinate conelets. Some lava from this fissure
infilled a nested crater in the northwestern part of Tagus Cone. (Photograph
by U.S. Air Force.)

connecting the central craters of Tagus and Beagle cones were prolonged, it would intersect the crater of the old Cape Berkeley volcano.

All three cones were formed by phreatomagmatic eruptions comparable to those that produced such well-known cones as those of Diamond Head and Koko Head, near Honolulu, Hawaii. The original vents of all three cones were probably submarine, and, even if not, they surely cut beds that were plentifully charged with sea water. The rising magma was therefore drastically chilled and blown out as fragments of basaltic glass (sideromelane), most of which has since been changed to palagonite. Cindery lapilli and rounded bombs are conspicuously lacking among the ejecta. These consist, on the contrary, of yellowish and brownish tuff, mainly of sand and gravel size, mixed with angular and subangular fragments of basalt and sporadic blocks of coarse-grained, plutonic rocks that are referred to below. None of the vents erupted lava. Countless explosions must have taken place in rapid succession, each thin layer of ejecta blanketing the preceding one to produce a conspicuous mantle bedding that conforms in general with the outer slopes of the cone and the inner slopes of the craters, as Darwin noted long ago. In what order the three cones were built we did not ascertain, but they must have developed in quick succession and perhaps partly at the same time. Until quite recently the cones were islands; they have now been linked to Darwin volcano by accumulation of youthful lavas along the coastal plain.

Tagus Cone has at least four nested craters. The youngest crater, which marks the central vent, contains a small salt lake enclosed by a low cone of tuff; the other craters mark peripheral vents (Pl. 10). Perhaps the oldest crater is the one breached to form Tagus Cove; a second crater lies a short distance east of the central one containing the salt lake; and a third one, partly infilled with recent lava, lies northwest of the lake.

The tuff of Tagus Cone contains much less lithic debris than does that of Beagle Cone, and most of this debris is smaller. Along the rim of the crater, much of the tuff is pisolitic, that is to say, it contains abundant, small accretionary lapilli formed when fine ash fell through clouds of steam and rain or through showers of spray during submarine eruptions.

The northern side of the cone is crossed by a chain of conelets built of scoria and agglutinate, most of them between 10 and 50 feet across; these are the "bubble craters" mentioned by Chubb. They lie on a radial fissure cutting the flank of Darwin volcano. In the depression between the base of Darwin volcano and Tagus Cone, a line of lava-scoria vents on the same fissure discharged copious flows of basalt. Lava also issued from the western end of the fissure to pour into the northwestern crater of Tagus Cone (Pl. 10).

The youngest flows immediately north of Tagus Cone, which were erupted from fissures on the flank of Darwin volcano, consist for the most part of broken, slabby pahoehoe lava dotted with many pressure ridges. Close to the coast, however, the medial portions of the flows are thinly mantled with "cindery" ejecta, products of phreatomagmatic explosions set off when the lavas spread over wet surfaces or poured into shallow, shoreline bays. Around Lake Myvatn in Iceland, and along the coasts of Hawaii, large cones of basaltic scoria have been produced where lavas poured into water; but here the explosions were not as violent and were shorter lived so that they formed only a veneer of fragmental debris.

Beagle Cone, unlike Tagus, was built by a single series of eruptions from a central crater. The southern wall of the crater has been breached, presumably by marine erosion, and youthful flows of basalt, erupted from fissures on the flank of Darwin volcano, have poured through the breach to impound a salt lake on the crater floor. The tuffs that compose the cone include many angular lapilli and blocks of basalt, some of them 4 feet in diameter, and, as Darwin first noted, they also contain many fragments of coarse-grained plutonic rocks. These coarsely crystalline fragments are clearly crystal cumulates from the basaltic magma.

Wolf Volcano

The northernmost and highest of the Albemarle shield volcanoes, now known as Wolf, was formerly called Mount Whiton. It reaches a height of 5600 feet. Its profile, like that of Narborough volcano, somewhat resembles that of an overturned soup plate, the flattish top and basal parts being separated by abnormally steep slopes, locally reaching angles of 35° (Pl. 11). The summit caldera measures roughly 4 miles by 3 at the rim and is elongated in a north-northwest direction, parallel to the major axis of Albemarle Island. Its depth approximates 2000 feet. Most of the caldera floor, as may be seen in the aerial photograph (Pl. 12), is covered by youthful lavas, some of which poured from fissures on the western wall and some from fissures on the opposite wall. A broad terrace on the western wall, evidence of peacemeal collapse, is traversed by an arcuate fissure into which lava has poured from the rim of the caldera. Arcuate fissures, marked by lines of small spatter and scoria cones, are especially numerous along and just beyond the southern and eastern rims of the caldera, and scores of lava flows have been discharged from them in very recent times to pour down to the sea. Radial fissures are also present on the flanks of the shield, particularly on the western, northern, and northeastern flanks (Pl. 13).

VOLCANIC PROFILES

Steep northeastern and eastern slopes of Wolf volcano, Albemarle Island. The profile resembles that of an overturned soup plate. (Photograph by Alden Miller.)

WOLF CALDERA, ALBEMARLE ISLAND

The bench on the western wall, about 400 feet above the floor, is crossed by a narrow fissure into which a recent lava has poured from vents along the southwestern rim of the caldera; other recent flows from these vents and from vents near the eastern rim have cascaded down to spread over the caldera floor. Many arcuate lines of scoria and spatter cones border the eastern rim of the caldera, and numerous flows from these cones have recently poured down the steep flanks of the shield to reach the sea. (Photograph by U.S. Air Force, 1946.)

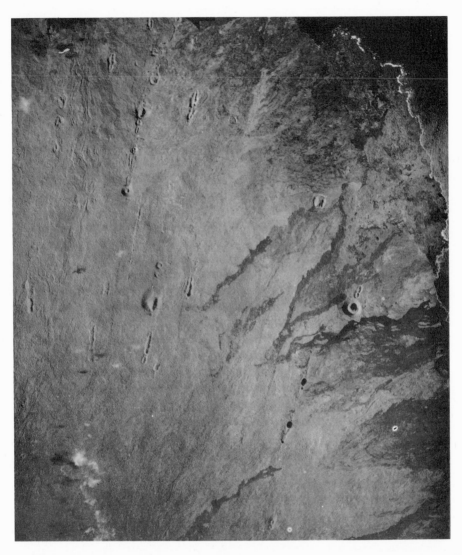

NORTHEASTERN SLOPES OF WOLF VOLCANO, ALBEMARLE ISLAND

Lines of small cones, many breached on their lower sides, and one line of collapse pits are shown. These lines mark fissures that are radial with respect to the summit caldera of Wolf volcano. (Photograph by U.S. Air Force, 1946.)

Richards (1962) lists eight reported eruptions of Wolf volcano, in 1797, 1800, 1925, 1933, 1935, 1938, 1948, and 1953. To this list must now be added an eruption in 1963. The number of unrecorded eruptions, even during historic times, must be very large.

An account of the last eruption was written for us by Dr. David Snow, who was then Director of the Darwin Station. It concerns a flank flow of aa lava.

The eruption started at the beginning of March 1963, but the exact date is not known. On March 4, as red glow was visible from Indefatigable (Santa Cruz) Island in the evening, pulsating approximately every 20 seconds. It was still visible as late as March 16. On March 28, I visited the east coast of Albemarle Island and found that a flow had reached the coast about a mile south of Cabo Marshall; it had pushed out about 50 yards into the sea over a front of about 200 yards. It was still hot and fumes were coming out of the cracks. The surface of the flow was composed of great fragments of ash and powder, tumbled boulders, loose rocks, etc., the whole thing loose and unstable. It did not differ from other recent flows along the same stretch of coast, and when cold could certainly not be distinguished from them. A little smoke was still coming from the cone on the southern flank of Volcan Wolf, on the skyline when viewed from the coast, at an estimated height of 2000 feet, some 4 miles from the coast.

Cape Berkeley Volcano

Having summarized the main features of the five shield volcanoes of Albemarle Island, we turn next to describe the remnant of a huge volcano near Cape Berkeley, at the northwestern end of the island (Fig. 17). Only about one-third of the original volcano remains, the rest having been downfaulted into the sea, leaving a line of precipitous cliffs. Even so, the remnant third rises to a height of more than 2600 feet, so the height of the original volcano must have been close to 3000 feet, despite the fact that its base was only about 4 miles across. No major volcano in the entire archipelago has steeper sides. Most of it seems to consist of basaltic lavas, but these are interbedded with many thick layers of coarse fragmental ejecta, and it is largely to their presence that we ascribe the steepness of the slopes.

Downfaulting of the western part of the volcano appears to have taken place along more or less arcuate faults. Aerial photographs reveal two parallel sets of faults separated by a bench close to the foot of the precipitous cliffs, the inner set marked by a line of small spatter and scoria cones from which cascades of aa lava descended to the coastal flats in quite recent times. Other youthful lavas, but of pahoehoe type, are widespread near Cape Berkeley itself, close to the equator, their source fissures being marked by chains of small agglutinate cones.

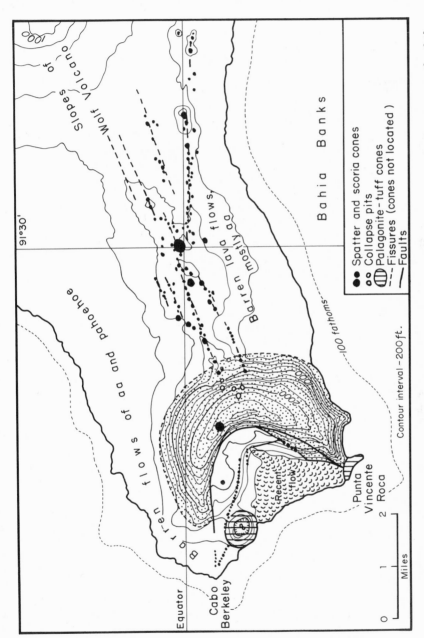

Figure 17. Cape Berkeley peninsula. Probably many more fissures and cones are in the area north of the equator, but aerial photographs were not examined.

Cape Berkeley volcano was formerly separated from Wolf volcano, and for a time it must have been an island, but now the two volcanoes are linked by a high ridge built mostly of aa lavas erupted from swarms of subparallel fissures along and near the crest (Fig. 17). Scores, if not hundreds, of spatter and scoria cones are aligned along these fissures, and innumerable flows have poured from them in recent times. Where the fissures abut against the side of Cape Berkeley volcano, there are at least six small collapse pits, and a short distance to the north, high on the flank of the old volcano, there is a young cone of scoria.

The arcuate faults at the base of the Cape Berkeley cliffs suggest that the main reason for the disappearance of the western part of the volcano may have been subterranean withdrawal of magma, possibly occasioned by submarine eruptions or landslides. After the collapse, two cones of well-bedded, yellowish and brownish sideromelane and palagonite tuff were built on the downdropped block, the smaller and more eroded of the two now forming Punta Vincente Roca.

Petrography

Only a few specimens were collected from Albemarle Island, and nearly all of these closely resemble lavas already described by Richardson (1933) and by Banfield and his colleagues (1956). However, the lavas of Cape Berkeley, not previously studied, differ from those of the large shield volcanoes and therefore merit special mention. Of three specimens from very recent lavas that reach the coast near Cape Berkeley, one has been analyzed chemically (Table 2a, No. 56) and is illustrated in Figure 18c. All are olivine basalts with rare microphenocrysts of sodic labradorite, olivine (Fa_{13}), and augite ($2V_z = 56°$). Their fine-grained, intergranular groundmass consists of sodic labradorite (40 percent), faintly zoned olivine partly altered to iddingsite (20 percent), clinopyroxene (30 percent), anorthoclase, magnetite, ilmenite, and hematite, with accessory apatite. A specimen from the large recent cone near the point of Cape Berkeley (No. 55), while having essentially the same mineral composition, has an unusual groundmass texture. Two kinds of plagioclase are present; one, sodic labradorite, occurs as slender laths; the other, probably potassic andesine, is more abundant and forms equidimensional, untwinned grains that grew ophitically around the labradorite and mafic minerals.

These Cape Berkeley lavas are the only olivine basalts we know of on Albemarle Island; all others seem to be less alkalic, generally with from 2 to 20 percent plagioclase phenocrysts in a matrix of labradorite laths, clinopyroxene, magnetite, and glass. Banfield and his colleagues remarked on the scarcity of olivine in the lavas of

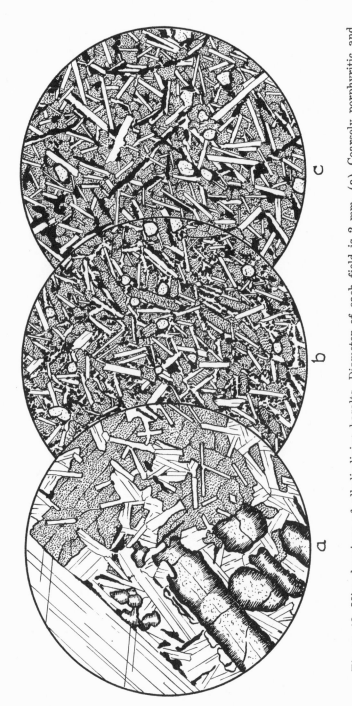

Figure 18. Microdrawings of alkali-olivine basalts. Diameter of each field is 2 mm. (a) Coarsely porphyritic and ophitic basalt from Wenman Island (No. 39). At the upper left is part of a phenocryst of bytownite 6 cm long; at right, a large ophitic plate of titaniferous augite; partly iddingsi-tized olivines; laths of labradorite, and plates of ilmenite. (b.) Nonporphyritic, intergranular basalt from southern end of Abingdon Island (No. 4, Table 2b). Laths of calcic andesine and sodic labradorite are separated by or, locally, are partly enveloped by purplish-brown augites, many of which have subhedral, prismatic forms. Ovoid granules of olivine and abundant ilmenite. (c.) Intergranular aa flow from recent eruption at Cape Berkeley, Albemarle Island (No. 56, Table 2a). Laths of labradorite, separated by granules of augite and fewer of olivine. Abundant plates of ilmenite. A little interstitial anorthoclase is present but is not shown in the drawing.

the principal volcanoes, and our observations support this general rule.

The coarse-grained lapilli and blocks among the ejecta of Tagus and Beagle cones were all described by Richardson. Most of them are eucrites made up of bytownite, olivine, pale-green or brownish augite, and interstitial glass, many having a porous, diktytaxitic texture resulting from drainage of liquid from the coarse crystal aggregates. Bytownite ($\sim$ An$_{80}$) accounts for 40 to 80 percent of the volume of the specimens we examined. Olivine (Fa$_{22}$) makes up between 10 and 20 percent. An analysis of a diopsidic augite from one inclusion is shown in Table 16 (No. 64) together with its optical properties. The relation of the inclusions to their host lavas is discussed on page 133.

A fine-grained basalt (Table 2a, No. 63), probably typical of the host rocks of the gabbroic inclusions, is a tholeiite with small, slender laths of medium labradorite and intergranular clinopyroxene ($2V_Z = 52°$), magnetite, ilmenite, and apatite.

A lava (No. 57) that poured between Tagus and Beagle cones from a source on the flank of Darwin volcano is a strongly porphyritic basalt containing about 35 percent xenocrysts of medium bytownite, up to 1 cm in length. Augite xenocrysts are smaller and less abundant, and xenocrysts of olivine are present only in very minor amount. All of these xenocrysts have shapes and optical properties similar to those of the coarse-grained crystal aggregates ejected from Tagus and Beagle cones, and they were probably derived from similar plutonic rocks. The enclosing lava is a fine-grained inter-growth of medium labradorite laths, intergranular augite, and abundant magnetite.

The trachyte pumice recently erupted by Alcedo volcano is extremely poor in crystals and all of these are very small. They consist almost entirely of plagioclase and a dark-green, nonpleochroic, hedenbergitic clinopyroxene (Z to c $= \sim 45°$; $N_Y = 1.718$); with these are rare prisms of greenish-brown hornblende. Neither quartz nor sanidine was detected. The pumice itself is remarkably inflated, the walls between the vesicles being extremely thin. The refractive index of the colorless glass is 1.513 ± 0.002. A chemical analysis of the trachyte is presented in Table 10 (No. 130), and the place of the rock in the scheme of magmatic differentiation is discussed on page 145.

ROCA REDONDA

The islet of Roca Redonda, which lies some 15 miles northwest of Albemarle Island, is only about 400 yards long and nowhere even half as wide. It is encircled by precipitous cliffs that rise to a flattish top at an elevation of about 220 feet (Pl. 14). The islet is merely the emergent top of an enormous volcano, 15 miles wide at the base, that rises from the ocean floor at a depth of almost 10,000 feet (Pl. 1). It lies in line with Darwin and Alcedo volcanoes on Albemarle Island and with a profound submarine depression some 25 miles to the northwest. The flattish top, which actually dips very gently to the southwest, consists of pahoehoe lava, and, from the air, the cliffs appear to be made up of similar lavas, their thickness ranging from about 2 to 10 feet. Nobody, as far as we know, has ever landed on the rock.

NARBOROUGH (FERNANDINA) ISLAND

Narborough volcano is one of the most active and surely one of the most impressive volcanoes in the world. But because of its high, steep, and rugged slopes it is very little known, and undoubtedly many unrecorded eruptions have taken place during historic times.

The entire island, measuring 18 by 15 miles (Pl. 15), is occupied by this one volcano. Below 1000 feet, the slopes are extremely gentle, but between 1000 and about 4000 feet they become abnormally steep; at still higher elevations the slopes again become very gentle, so that a wide, flattish bench surrounds the summit caldera. The caldera itself is an awesome, steep-walled pit, between 2500 and 2750 feet deep, measuring approximately 4 miles by 2.75 miles across at the rim and 2 miles by about 1.5 across the floor (Pl. 16).

The unusual form of the volcano may be judged from the cross sections (Fig. 19) and the sketch (Fig. 20). Seen from the sea, the profile resembles that of an overturned soup plate. Mokuaweoweo, the large caldera on top of the Hawaiian shield volcano of Mauna Loa, is also surrounded by a wide, flattish bench, but beyond this its slopes descend at angles of between 3° and 6°, whereas those of the Narborough volcano descend at angles ranging from 15° to 34°. Some of the great shield volcanoes of Albemarle Island, particularly Wolf volcano, closely resemble the Narborough shield, but we know of no other large shield volcanoes in the world with slopes as steep. As to the origin of these steep slopes we remain in doubt. They may have been formed in part by injection of sills from the central conduit, but in the main they probably formed by lateral distension of the volcano when ring dikes were injected into the swarm of concentric fissures that surround the caldera. Both of these processes would cause an upward and outward swelling of the entire volcanic edifice. As to the very gentle slopes close to the base of the volcano, there is much less doubt; they reflect the fact that most of the lavas around the base were poured out in a fluid condition from fissures far down the flanks.

ROCA REDONDA

McBIRNEY AND WILLIAMS, PLATE 14
Geological Society of America Memoir 118

NARBOROUGH (FERNANDINA) CALDERA

Note the straight and arcuate fissures, marked by lines of scoria and spatter cones, beyond the rim of the caldera. Almost all of the recent lavas from these fissures have poured down the steep, outer flanks of the shield, but a few have poured into the caldera, for example, on the northwestern side. Benches on the caldera walls testify to piecemeal collapses. Several small scoria cones are to be seen on the caldera walls, particularly on the south-southwestern side, where they are arranged in arcuate lines. When this photograph was taken, in May 1946, the floor of the caldera was occupied mostly by a large lake surrounding a big scoria cone. In 1957 or 1958, the lake was almost obliterated by an extensive lava flow, so it is now restricted to the northern end of the caldera floor. (Photograph by U.S. Air Force.)

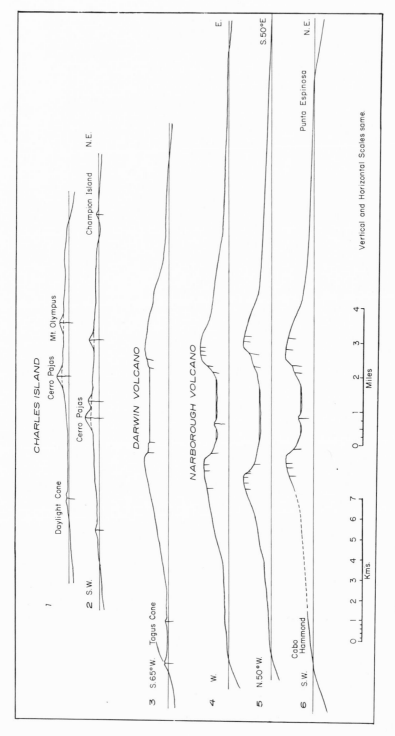

Figure 19. Cross sections of Charles Island and of the Darwin and Narborough volcanoes.

Figure 20. Narborough (Fernandina) volcano. View from Punta Espinosa, looking southwestward. Note that the volcano has the form of an inverted soup plate. The hidden summit caldera is surrounded by a broad, flattish bench; abundant youthful flows descend the steep, outer slopes. Recent pahoehoe flows in the foreground.

Neither on Narborough volcano nor on any of the five shield volcanoes of Albemarle Island are there any large parasitic cones of scoria comparable with those on Charles, Chatham, James, Indefatigable, and Bindloe islands. However, small cones of scoria, agglutinate, and spatter are abundant, as they are on the other islands. This lack of large pyroclastic cones probably indicates that the great shields of Narborough and Albemarle are at present in the mature stage of rapid growth, like Mauna Loa and Kilauea on Hawaii.

The principal products of Hawaiian volcanoes during their maturity are thin, wide-spreading flows of tholeiitic basalt, accompanied by small cones of scoria and spatter built along eruptive fissures. During later stages of growth, thicker flows tend to accumulate and ultimately to fill the summit caldera, their composition changing gradually from tholeiitic to alkali-olivine basalts. Still later, during the "Mauna Kea stage" of evolution, more varied types of material are erupted, and, in general, explosive activity becomes more pronounced while intervals between eruptions become longer (Macdonald and Katsura, 1962). This sequence of eruptive activity and the resultant morphological changes seem to apply also to many of the major volcanoes of the Galápagos archipelago, and, to a lesser degree, so does the petrographic evolution.

One of the most arresting features of the Narborough volcano is the swarm of arcuate, circumferential fissures traversing the wide, flattish bench around the summit caldera (Pl. 17). We know of no

NORTHWESTERN END OF THE NARBOROUGH (FERNANDINA)
CALDERA

View looking northeastward. Pronounced bench left by collapse and a very
recent flow that cascaded from the caldera rim, inundated part of the bench, and
continued to the caldera floor. (Photograph by David Cavagnaro, February 1964.)

volcano in the world that displays this feature more vividly. Around part of the caldera, there are at least four concentric sets of fissures, the outermost a mile or so from the rim; elsewhere only a single fissure or two fissures can be seen. Some fissures are open cracks that show no signs of vertical displacement of the walls; most, however, are marked by rows of scoria and spatter cones. Innumerable flows of lava have poured from these fissures during recent times, but very few have poured into the caldera, even from fissures no more than 100 to 200 yards from the rim; all other flows have poured in the opposite direction, down the steep, outer slopes of the shield. Arcuate fissures are also present on the walls of the caldera, as well as arcuate faults, as the aerial photograph (Pl. 16) shows. A few small cones of scoria and spatter have been built along these intracaldera fissures, and some lavas have issued from them to cascade to the caldera floor.

Radial fissures, so conspicuous on the shield volcanoes of Albemarle Island, are relatively rare on Narborough, although the photograph (Pl. 17) shows one arcuate fissure near the rim of the caldera that swerves at its northern end to become a radial fissure. Swarms of straight, subparallel fissures, similar to those linking the Albemarle volcanoes, are apparently absent from the Narborough volcano; certainly none were seen cutting the walls of the caldera.

All the lavas that we saw on the steep southern and southwestern slopes of the volcano are of aa type, and most of these issued from the arcuate fissures near the summit. A few vents are to be seen on the steep slopes themselves, but there are many more near the base. On the southeastern slopes, there are fewer flows from the arcuate summit fissures and most of them fail to reach the bottom; apparently most of the lavas on these slopes were discharged from flank fissures. On the northeastern slopes, between 1000 and 2000 feet, there are many small, parasitic cones; on the eastern slopes such cones are less numerous, although some are notably large. On all of these slopes, the lavas are of aa type. Around Punta Espinosa, however, there are extensive sheets of fresh pahoehoe lava that seem to have issued from vents close to the bottom of the steep slopes of the volcano.

Richards (1962) has summarized the recorded activity on Narborough Island since 1813. During this time there have been at least ten eruptions, most of them from the arcuate fissures near the summit and from parasitic vents on the flanks. In 1946, as may be seen in Plate 16, the floor of the caldera was occupied by a large lake, a cone of scoria, and a youthful lava flow, and this was how it appeared when Dr. Robert Bowman was there in 1957. In 1958, however, the lake had disappeared, and a group of Norwegians reported that the caldera floor was too hot to cross on foot. When we flew over the caldera in February 1964, there was only a small lake, confined to the northern part of the floor, the remainder being covered by very fresh lava. It seems likely that this lava was erupted in 1957 or 1958, between the visits of Bowman and the Norwegians.

Collapse of the Caldera in 1968

Simkin and Howard (1968) have described the remarkable activity of June 1968 when the floor of the caldera suddenly collapsed. On May 21 there had been a brief eruption at an elevation of about 700 m on the eastern flank of the volcano, and lava covered about 10 km². The eruption seems to have lasted less than a day. At 1018 on June 11 an earthquake near Narborough was recorded in Quito, Ecuador. Within an hour, a large white cloud had risen high enough to be visible at Academy Bay, 140 km to the east. The cloud rose as a column, then spread symmetrically at a height of about 20 km. At about 1600, a dark ash cloud appeared and members of the crew of the fishing boat *San Jose,* which was in Urvina Bay, 35 km east of the caldera, reported a pink color on the underside. The cloud grew and spread rapidly over a wide region.

A loud boom at 1708 was heard 220 km to the east at Wreck Bay. This was followed by a series of smaller explosions that continued at intervals of 1 to 5 minutes for about 2 hours. Infrasonic long-wave shocks resulting from these explosions were recorded throughout the western hemisphere. Their amplitude measured at Boulder, Colorado, was comparable to that of the largest nuclear explosions.

Lightning was observed over the volcano for about 4 hours beginning at 1900, and flashes of red, green, and violet light, possibly flames of burning gas, were seen from the *San Jose.* About the same time, ash began to fall over much of the western part of the archipelago.

An earthquake at 1720 was felt throughout the islands. More shocks followed, an average of about 5 a day being recorded in Quito from June 12 to 14. The number of seismic events increased to a maximum on June 18 and 19, when the seismograph at Academy Bay recorded about 200 a day. A scientific party from the Charles Darwin Research Station was on the flank of the volcano on June 19 and counted 56 tremors in a 6-hour period. On the same day there were 55 events with a magnitude greater than 4.0. The largest shocks were on the 15th when there were four with a magnitude of over 5.

The party from the Darwin Station, which reached the summit of Narborough at 1330 on June 19, was the first to view the caldera after the beginning of the eruption. At that time the interior was completely obscured by the dust of constant avalanches. They detected no evidence of new lava. On the morning of the 21st, during a lull in the avalanching, Capt. F. E. Sevilla of the Ecuadorian Air Force flew over the volcano and was able to see the floor of the caldera. The small lake was still visible but had shifted to the southeast. No fresh lava was seen, although there were small jets of white steam on the northwestern part of the floor. Simkin and Howard visited the caldera between July 10 and 13 and found no further change. Photographs of the caldera taken at that time can be compared with those taken shortly before the collapse (Pl. 18).

The floor of the caldera subsided asymmetrically, the southeastern part dropping more than 300 m while only minor subsidence occurred on the opposite side. The total volume of the collapse was between 1 and 2 km^3. Movement took place along an elliptical fault that bounded the caldera floor. The same fault probably controlled previous subsidences. The northern third of the floor was broken by numerous transverse fractures. A small block, about 0.5 km^2 in area on the western side, dropped independently and about one-third of it is now 50 to 150 m below the main floor level. This same block seems to have been separate from the main block before the collapse, also, because it formerly formed a terrace about 100 m above the floor. Apart from these features and minor sagging near the center, most of the floor fell as a coherent block.

Fumarolic activity was concentrated near the center and along the western and northern edges of the smaller block. Increased fumarolic activity had already been noticed in this area by a party that had climbed the volcano the previous February. The most intense activity in mid-July was centered in a new low-rimmed explosion crater near the center of the block. Simkin and Howard surmised that this crater, which they estimated to have a diameter of about 50 m, was formed during the activity of June 11, but they think it is too small to be the sole source of the ash erupted at that time.

New fractures formed around the caldera rim, mainly at the edge and becoming progressively fewer outward for a distance of about 500 m. Most were simple tension fractures with little vertical displacement.

Fragmental ejecta seem to have been entirely composed of lithic debris torn from the walls of the vent. The total volume was between 10^7 and 10^9 kg, equivalent to only a few thousand cubic meters of solid rock. These ejecta, together with the lava discharged by the flank eruption just before the collapse, account for less than 10 percent of the engulfed volume. Even if additional new lava should be found on the floor of the caldera, it is doubtful that it will approach the volume needed to explain the collapse. The deficiency might be explained if there were evidence for a large intrusion, such as that which seems to have caused the elevation of Urvina Bay in 1954,

CHANGES IN NARBOROUGH CALDERA AS A RESULT OF THE COLLAPSE OF JUNE 1968

Panoramic view of Narborough caldera before and after the collapse of June 1968. Figure 1 was taken by Dr. P. Colinvaux from the northeastern rim in 1966. Figure 2 was taken by Dr. Tom Simkin in July 1968 from a position somewhat north and west of Figure 1. Figure 2 spans about 150° and includes a wider extent of the summit region than Figure 1, but the two photographs are approximately equal in scale and illustrate the extent of the collapse. Note the change in position of the lake from the northwestern to the southeastern end of the caldera.

but Simkin and Howard report no changes of the shoreline around the island. It must be concluded that collapse was caused by withdrawal of magma at some unknown depth, possibly by an outbreak of lava at the base of the escarpment of the Galápagos Platform just west of the island.

Petrography

All the specimens we collected from Narborough volcano are essentially alike. We examined two lavas from elevations of approximately 1200 (No. 43) and 1800 feet (No. 42) on the southern slopes. Phenocrysts and glomerocrysts of sodic bytownite total about 8 percent of the volume; some are as much as 4 mm long, but most measure less than 1 mm long. Microphenocrysts of olivine (Fa_{13}), present in the amount of less than 1 percent, show no detectable reaction relations with the groundmass minerals. The groundmass itself has an intergranular texture, consisting of laths of medium to calcic labradorite, augite ($2V_z = 56°$), and vesicular glass rich in magnetite. One specimen (No. 42) contains rare grains of pigeonite in the groundmass; the other (No. 43) contains a few of hypersthene. Analyses of one of these specimens (No. 42) is presented in Table 2a.

A specimen of pahoehoe lava from Punta Espinosa contains about 4 percent glomerocrysts, up to 2 mm long, of calcic labradorite with thin sodic rims and less than 1 percent of olivine and augite phenocrysts. The groundmass has an intersertal texture and contains more glass than do the lavas from the slopes of the volcano. An analysis of this lava (Table 2a, No. 87) shows it to be a tholeiitic basalt with 1.36 percent normative quartz.

TOWER (GENOVESA) ISLAND

Almost nothing was known about the geology of this small, low island until Professor J. C. Granja went there in January 1964 as a member of the Galápagos International Scientific Project. The bathymetric map (Pl. 1) shows that the island is the highest point on a long submarine ridge that extends east-northeast, following one of the two dominant fracture systems of the archipelago. The slopes of this submarine ridge descend to depths of approximately 800 fathoms, but the island itself rises to a height of only about 210 feet, although it measures 2.5 miles across (Fig. 21).

The island is almost equidimensional in plan, but it is indented on the southern side by a shallow, almost circular inlet called Darwin Bay. The main vent of the volcano lies close to the center of the island. It is a circular crater, approximately 2000 feet wide at the rim and 200 feet deep, with a lake 1150 feet across on its floor. From this central vent most of the slopes descend gradually and un-interruptedly to the coast, but locally they are cut off by cliffs up to 100 feet in height. The island, according to Professor Granja (1964), is largely covered with youthful flows of basaltic lava, predominantly of pahoehoe type, supporting only a very sparse growth of *Opuntia* cacti. In his opinion, the island was form-erly a single shield volcano with a summit crater; then, after a period of quiet, activity was re-sumed in recent times, not from the central vent but from fissures on the flanks. But even the shield must be fairly young, as its topo-graphic form suggests.

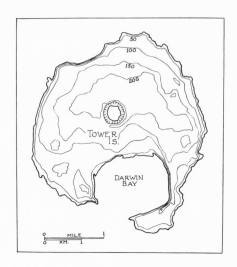

Figure 21. Map of Tower (Geno-vesa) Island. Contour interval, 50 feet. (*After* U.S. Hydrographic Office chart No. 5945, 1947.)

Many fissures cut the younger lavas, some of considerable length and depth. One fissure, between the central crater and Darwin Bay, trends N.70°W., parallel to the submarine ridge already mentioned; other fissures crossing the northern part of the island trend north-eastward. In places, the older lavas have been uparched by the rise of younger ones, and many small driblet cones (hornitos) developed where the younger lavas emerged from tension cracks. Pressure ridges and lava tubes are plentiful, and on the southeastern side of the island, where rising lava came in contact with sea water, brown-ish, scoriaceous, pumicelike material was formed, rich in large pheno-crysts of plagioclase.

Professor Wyatt Durham informs us that waterworn lumps of white pumice are plentiful on the beaches of the island, but their source remains in doubt.

Petrography

The only specimen we examined was one collected by Professor Granja (No. 92). This lava, which is illustrated in Figure 6c, is a strongly porphyritic basalt similar to those from Abingdon, Bindloe, Culpepper, and Wenman islands. About 40 percent of the volume consists of phenocrysts of plagioclase up to 2 cm in length. Their composition is rather uniform medium bytownite, but most crystals have thin rims of labradorite. Olivine phenocrysts are slightly more abundant and larger than in the specimens from Abingdon Island,

and they total between 1 and 2 percent of the volume. Micropheno-crysts of clinopyroxene are also present but are even less numerous. Both mafic minerals are found almost exclusively in interstices be-tween larger plagioclase crystals. The coarse-grained, intergranular to subophitic groundmass is made up of medium labradorite (30 percent), clinopyroxene (40 percent), reddened olivine (20 percent), and titaniferous magnetite (10 percent). Apatite and analcite are accessories.

According to Professor Granja, there are also nonporphyritic basalts on Tower Island, but we did not have an opportunity to study them.

BINDLOE (MARCHENA) ISLAND

The desolate island of Bindloe (Fig. 22) seems to be the summit of a large shield volcano in the "Mauna Kea stage" of development. It seems to consist, in other words, of a cluster of pyroclastic cones

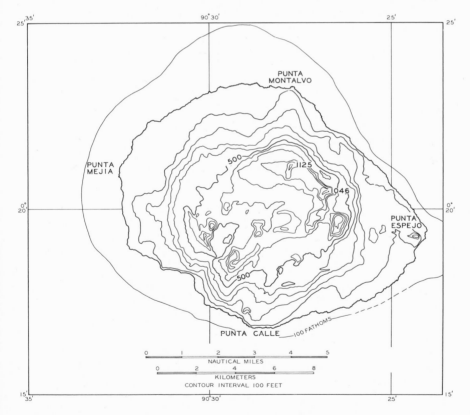

Figure 22. Map of Bindloe (Marchena) Island. Note that the distribution of peaks and ridges in the upper part suggests the outline of a buried caldera. (*After* U.S. Hydrographic Office chart No. 5945, 1st ed., 1947.)

and lava flows that have buried the summit caldera and surrounding slopes of an earlier shield volcano of "Mauna Loa type." Richards and Chubb called attention to what they thought was a caldera on the island, influenced no doubt by the presence of an arcuate ridge that forms the highest part; however, the supposed caldera has been completely obliterated by the products of later eruptions, and only the arrangement of some of the eroded tuff cones suggests its former presence. None of the older lavas of the volcano, those erupted during the "Mauna Loa stage," seems to be exposed.

Our stay on Bindloe Island was limited to part of a day on the southwestern coast and a brief helicopter flight to the summit area the next morning. A line of cliffs, 10 to 15 feet high, runs along the southwestern coast for more than 2 miles. These cliffs are cut in well-stratified, mantle-bedded, buff-colored sideromelane and palagonite tuffs that dip seaward at about 10° (Pl. 19, fig. 1). Cliffs of similar tuff were seen from the air, on the outer side of the arcuate ridge that runs through the topmost peaks of the island, and perhaps the entire ridge, as well as many of the eroded, pale-colored hills in the summit area, consist of similar ejecta. Some hills have crescentic outlines, suggesting that they are the remains of individual cones; others are quite irregular in outline (Pl. 19, fig. 2).

Floods of barren, black pahoehoe lava have completely buried the supposed caldera, which, judging by the topography, may have measured almost 5 miles in an east-northeast direction and 3 miles across. These youthful lavas partly inundated the ridges and cones of tuff, and many of them cascaded down the outer slopes of the volcano to reach the sea. Two measurements taken by Allan Cox along the southwestern coast showed the lavas to have normal magnetization.

The youngest flows on the island are probably less than a thousand years old; perhaps some were erupted within the last few centuries. Slevin (1931) reported that when he was on the island in 1906 there were fumaroles on the northeastern peak, and Richards (1962) cites a report of Raymond Lévêque that an islander saw "smoke" rising from the higher parts of the island during recent years. Miguel Castro, a keen observer and one of the best informed inhabitants of the archipelago, told us that sometimes, especially on cold mornings, plumes of steam can be seen rising from fissures on the southwestern slopes of the island.

If we are correct in supposing that most of the fragmental ejecta on the island are sideromelane and palagonite tuffs rather than basaltic scoria, the principal explosive activity must have been of phreatomagmatic type, and this in turn suggests that when the explosions took place the inferred caldera may have contained a lake or at least was surrounded and underlain by abundant ground water infiltrating from the sea.

Figure 1. Sideromelane and palagonite tuffs on the southwestern coast.

Figure 2. Part of the summit area, perhaps overlying a buried caldera. Shows a group of tuff cones partly inundated by floods of Holocene pahoehoe lava.

BINDLOE (MARCHENA) ISLAND

McBIRNEY AND WILLIAMS, PLATE 19
Geological Society of America Memoir 118

Petrography

Most of the lavas of Bindloe Island resemble the nonporphyritic alkali-olivine basalts of neighboring Abingdon Island. They contain only a very few phenocrysts, up to 3 mm across, of clear, faintly zoned, sodic labradorite with thin rims of andesine. Microphenocrysts of clinopyroxene and olivine amount to less than 1 percent of the volume. The groundmasses of the specimens we examined contain little glass; they are composed almost wholly of calcic andesine, clinopyroxene, reddened olivine, anorthoclase, ilmenite, and titaniferous magnetite. Analcite is present in the vesicles of almost every section.

A specimen of porphyritic lava from the southwestern shore of the island resembles the strongly porphyritic lavas of Abingdon Island described below. It carries abundant phenocrysts of plagioclase, up to 1.5 cm long, and rare microphenocrysts of olivine. The groundmass is made up of sodic labradorite, clinopyroxene, olivine, ilmenite, magnetite, and minor amounts of anorthoclase and analcite.

Of particular interest is the fact that a few of the lavas that flooded the supposed caldera are tholeiitic. A specimen from the crust of a very fresh pahoehoe flow consists of small laths of medium labradorite enmeshed in a fine-grained intergrowth of clinopyroxene and magnetite. A minute amount of olivine may also be present in the groundmass, but our identification is not positive. Some of the clinopyroxene occurs as small, pale-brown, equidimensional crystals, but most of it forms sheaves and radiating clusters of thin blades, habits that seem to have resulted from rapid crystallization during the final stages of consolidation. A chemical analysis of this specimen is presented in Table 2b (No. 15).

Occasional small lumps of waterworn pumice are to be found on the shores of Bindloe Island. One that we examined (No. 91) contains a few minute crystals of calcic oligoclase and still fewer of sanidine and of a deep-green hedenbergitic pyroxene ($2V_z = \sim 55°$; Z to c $= \sim 45°$), along with needles of apatite. These minerals lie in a matrix of frothy, colorless glass having a refractive index of approximately 1.512. These observations suggest that the pumice is almost identical with that erupted by Alcedo volcano on Albemarle Island.

ABINGDON (PINTA) ISLAND

Abingdon Island, like Bindloe, rises from a high, northwest-trending submarine ridge. The sea floor immediately to the west drops abruptly to a depth of 1000 fathoms in a distance of slightly less than 5 miles and continues to fall in that direction until it reaches a depth of 1600 fathoms approximately 20 miles from the island. This conspicuous submarine scarp marks a major zone of faulting.

The island itself is elongated in the same direction as the submarine ridge on which it stands. It consists of two quite distinct parts (Fig. 23). The older part, which forms a narrow strip bordering the western coast, represents the remnant of a large volcano whose western part has been downfaulted into the sea, the high and

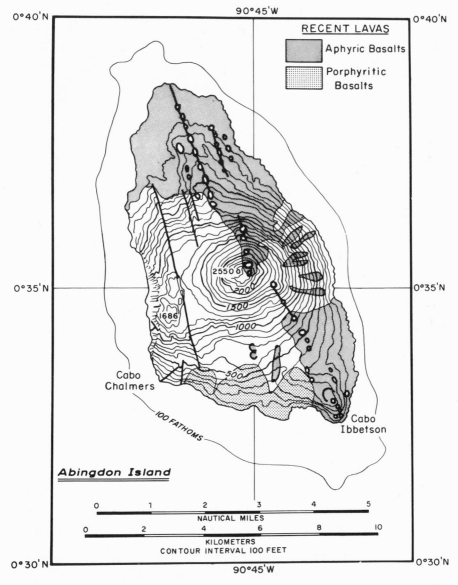

Figure 23. Map of Abingdon (Pinta) Island. (*After* U.S. Hydrographic Office chart No. 5945, 1st ed., 1947.)

precipitous coastal cliffs being an eroded fault scarp. A parallel fault inland separates this older part of the island from the main part that is occupied by a much younger volcano scarcely modified by erosion.

We examined the older, ruined volcano only from the air, but, as the photograph (Pl. 20, fig. 1) shows, it is made up of a thick succession of lavas, more than 30 flows or flow units being exposed in cliffs approximately 1300 feet in height. Locally, these lavas are cut by a few thin, steeply dipping and vertical dikes.

The younger volcano rises to a height of 2550 feet and had already reached this height, principally by eruptions from a summit vent, when the latest eruptions began from swarms of fissures that trend north-northwestward across its flanks. Slevin (1931) reported that when he was on the island in 1906 he saw vapors rising "from a hole in lava at the base of two small cones located on the northeast side of the summit," but when Dr. Robert Bowman visited the island in 1957 he found no signs of fumaroles near the summit.

There appear to be no records of historic eruptions on Abingdon Island; nevertheless, one cannot see the immense sheets of virtually barren, black lava and the perfectly preserved spatter and scoria cones from which they were discharged without concluding that there must have been many eruptions during the last few thousand years, if not during the last few centuries. Some idea of the youthful aspect of the landscape is given by the photograph (Pl. 20, fig. 1). Long chains of fresh cones and conelets vividly mark the eruptive fissures.

Most of the lavas, particularly at the northern end of the island, are of pahoehoe type, and generally they are dense basalts relieved only by a few phenocrysts of olivine. Parts of some flows, however, are extraordinarily rich in large phenocrysts of plagioclase (Pl. 21). In these strongly porphyritic parts, phenocrysts may constitute more than half the total volume, many of the crystals exceeding 2 cm in length. Yet these porphyritic parts pass abruptly into parts virtually devoid of phenocrysts, their boundaries crossing the ropy crusts of the flows at random, as may be seen in Plate 21, figure 1. We could not detect any systematic distribution of the two types of lava within individual flows; the strongly porphyritic parts may be found at the tops and bottoms of flows and scattered haphazardly throughout. Hence, a fairly clear separation of crystal-poor and crystal-rich parts must already have taken place within the magma chambers or in the feeding conduits before the lavas were erupted. The groundmass of the porphyritic lava is a markedly alkali basalt identical in composition with the adjacent nonporphyritic lava.

All the parasitic cones that lie inland are composed of dark basaltic scoria, but the large cone close to Cape Ibbetsen, at the southern tip of the island, in common with coastal cones on other islands, consists of sideromelane and palagonite tuff. It contains

Figure 1. Field of Holocene lava and line of youthful scoria cones at the southeastern end of the island.

Figure 2. Cliffed western coast, showing thin lava flows of the old volcano.

ABINGDON (PINTA) ISLAND

McBIRNEY AND WILLIAMS, PLATE 20
Geological Society of America Memoir 118

Figure 2. Hand specimen of the strongly porphyritic lava (scale in inches).

Figure 1. Holocene pahoehoe lava near the southern coast. Sharp contrast between irregular patches of strongly porphyritic and nonporphyritic basalt is shown.

ABINGDON ISLAND

many angular fragments of nonporphyritic basalt, but we found none of plutonic rocks.

Petrography

The lavas we examined from Abingdon Island are all alkali basalts. The nonporphyritic types, which are the most abundant, are fine-grained, vesicular basalts with an intersertal texture. Fully one-half of their volume is made up of slender laths of sodic labradorite. Clinopyroxene and olivine together account for nearly all the remainder, some olivine grains containing minute inclusions of spinel. Between the crystals are irregular patches of opaque glass. Analcite lines many vesicles, and in a specimen from the northeastern end of the island there are interstitial patches of an unidentifed zeolite.

Porphyritic lavas are widespread on the low southeastern slopes of the island. From 60 to 80 percent of their volume may consist of plagioclase phenocrysts up to 2 cm long. The plagioclase is generally calcic bytownite, but most crystals are normally zoned with rims of sodic labradorite. The groundmass plagioclase is present in two forms in approximately equal amounts, namely as slender, unzoned laths of calcic labradorite, about 0.5 mm long, and as xenomorphic crystals strongly zoned from calcic labradorite to oligoclase. Many of the unzoned laths are enclosed ophitically within clinopyroxene, whereas the xenomorphic crystals are more commonly interstitial. Olivine ($2V = \sim 90°$) amounts to only about 0.2 percent of the rock. It occurs in minute euhedral granules clustered around the rims of plagioclase phenocrysts. Pale-reddish-brown augite ($2V_z = 56°$) rarely reaches dimensions of even 0.1 mm. Some grains form subophitic growths around plagioclase; others occur in feathery bundles of slender blades. Ilmenite is abundant as thin plates and as chains of granules surrounding olivine and in the interstices between grains of clinopyroxene. Small amounts of anorthoclase, apatite, glass, and hematite make up the remainder of the rock.

Chemical analyses of the porphyritic basalt and of its groundmass are presented in Table 2b (Nos. 5, 5g). The normative compositions of both the total rock and its groundmass include more than 2 percent nepheline, indicating that there is approximately as much nepheline in solution in the plagioclase phenocrysts as there is in the groundmass, even though the mineral was not detected in the mode. The percentage of modal olivine is notably lower than that of normative olivine, although the mineral is more plentiful in other specimens that have a holocrystalline groundmass.

A specimen of nonporphyritic basalt from the same flow resembles other nonporphyritic basalts of the island, consisting of calcic andesine or sodic labradorite (about 50 percent), purplish-brown augite ($2V_z = 55.5°$; 30 percent), olivine ($2V = \sim90°$; 15 percent), ilmenite, and accessory zeolites (Fig. 18b). Its chemical composition (Table 2b, No. 4) is nearly identical to that of the groundmass of the porphyritic basalt (No. 5g).

CULPEPPER (DARWIN) AND WENMAN (WOLF) ISLANDS

The northernmost islands of the archipelago, Culpepper and Wenman, differ from all others in being detached from the Galápagos Platform. They are eroded tops of two huge volcanoes that rise from depths of more than 1000 fathoms, and they lie in line with the base of the great submarine fault scarp west of Bindloe and Abingdon islands. Two submarine volcanoes lie on this same line, one of them 30 and the other 45 miles southeast of Wenman Island, the former 3000 feet high and the latter more than 5000 feet high.

Culpepper Island is a small, flat-topped island about 550 feet high, encircled by unscalable cliffs. Allan Cox was one of the group to make the first helicopter landing. He informs us that the top of the island consists of basaltic tuff containing sporadic blocks up to 2 feet across. Beneath this is an essentially horizontal flow of porphyritic basalt, underlain in turn by a sheet of nonporphyritic basalt about 1 foot thick. Then follows a thicker layer of basalt crowded with phenocrysts of plagioclase, possibly the lower part of the nonporphyritic flow. A single magnetic measurement showed normal orientation.

Wenman Island is 830 feet high and is closely adjoined by three much smaller, lower islets (Fig. 24). Allan Cox and David Cavagnaro tell us that "large feldspar basalts" are widespread over most of the peninsula that makes up the northern end of the main island and are also plentiful at the opposite end. The prevailing dip of the flows is toward the south at low angles. Remanent magnetization of two lavas tested by Cox is reversed.

Figure 24 shows that the main island of the Wenman group has the shape of the letter I, the northern arm of which (in the foreground of the sketch) consists of two parts. The eastern part (on the left) consists of a pile of thin flows dipping southeastward; the western part, on the other hand, as well as the flat-topped islet in the foreground, consists of thicker flows that lie horizontally or almost so. Dr. Cox suggests that the thick, flat-lying lavas may represent part of the filling of the summit caldera of a large shield volcano built of thinner flows. If this is the correct interpretation, as we believe it to be, the relations resemble those observed on many eroded volcanoes in the Hawaiian Islands where flattish intracaldera lavas are surrounded by thinner, outwardly dipping flows.

Figure 24. Wenman (Wolf) Island viewed from the north. Near the center of the sketch are thin lava flows dipping gently to the east (left) and abutting abruptly against thicker, flat-lying lavas that may have accumulated within a caldera.

According to this hypothesis, the center of the original Wenman caldera must lie a short distance to the west of the eroded remnants.

Petrography

The lavas of Culpepper and Wenman islands, like those on all the islands in the northern part of the archipelago, fall into two types, namely, those with abundant plagioclase phenocrysts and those with none or virtually none. A typical nonporphyritic specimen from the northwestern corner of Culpepper Island is included in Table 2b (No. 35). About one-half of its volume is made up of laths of unzoned, sodic labradorite up to 1 mm long. Intergranular augite (~25 percent), olivine (20 percent), and titaniferous magnetite are the only mafic constituents. Analcite is common in vesicles and other cavities; natrolite and apatite are other accessories.

A specimen from the top of the cliffs at the northern tip of Wenman Island (No. 39) is typical of the porphyritic lavas and is illustrated in Figure 18a. Phenocrysts of sodic bytownite, with thin rims of sodic labradorite, reach 1.5 mm in length, and they make up at least one-half of the volume. The groundmass consists of sodic labradorite (40 percent), purplish-brown ophitic augite (30 percent), and olivine, partly altered to iddingsite (20 percent). Ilmenite and titaniferous magnetite account for the remaining 10 percent of the groundmass.

Pattern of Fractures and Faults

Two sets of fractures and faults can be distinguished in the Galápagos Islands, the first of a regional, tectonic nature and the second of a more local, volcanogenic nature related to the swelling and sinking of volcanic structures.

REGIONAL PATTERNS

Darwin was the first to point out that many Galápagos volcanoes are arranged in a more or less rectilinear manner, some along lines that trend north-northwest and others along lines at right angles thereto. This pattern is most graphically displayed on land by the major volcanoes of the J-shaped Albemarle Island (Fig. 16). Reference to the bathymetric map (Pl. 1) shows that these two trends are also recognizable on the sea floor, especially in the great submarine scarp west of Abingdon and Bindloe islands, in the submarine elongation of Tower Island, and in the alignment of three submerged volcanoes to the south. These "Darwinian trends" bisect the angle between the east-west Galápagos Fracture Zone and the Carnegie Ridge on the one hand and the northwest-southeast alignment of the Cocos Ridge on the other.

Several other lineaments are discernable among the Galápagos volcanoes, trending at slight angles to these "Darwinian trends." A northwest-southeast alignment is clearly revealed by Wenman and Culpepper islands and by two submarine volcanoes not far away and also by the elongation of the main volcano of James Island and the lines of small cones on its flanks. A north-northeast alignment can be seen in only a very few places, as in the vicinity of Daylight Point on Charles Island (Fig. 8). Approximately east-west alignments, parallel to the trend of the Galápagos Platform and Carnegie Ridge, are shown by several of the faults that cut the uplifted submarine lavas on Barrington, Baltra, and Indefatigable islands and by the belt of cones crossing the summit region of Indefatigable Island. No Galápagos volcanoes or faults are aligned in a north-south direction.

97

Rectilinear arrangements of volcanoes are uncommon elsewhere in the world. Such an arrangement may be discerned faintly among the volcanoes of the Kamchatka Peninsula, and, more clearly, among those of the Marquesas Islands. Chubb supposed that this arrangement of the Marquesas and Galápagos volcanoes was related genetically to an arcuate arrangement of the volcanoes of the Society and neighboring islands, far to the southwest. These South Sea volcanoes, in his opinion, rose from the crests of submarine folds produced by forces directed northeastward. These same forces, he thought, acted on a more or less rigid block of the ocean floor, the so-called Albatross Plateau, to produce the rectilinear features over which the Marquesas and Galápagos volcanoes were built. Later studies have shown, however, that the volcanoes of the Society and neighboring islands do not rise from the crests of submarine folds and that the fracture patterns in the Marquesas and Galápagos archipelagoes are much more complex than previously imagined.

Tectonic explanations must relate the fracture patterns within the Galápagos Islands to the east-west trends of the Galápagos Platform and Carnegie Ridge and of the Galápagos Fracture Zone, north of the archipelago and to the northeast-trend of the Cocos Ridge. All of these features must be considered in relation to the East Pacific Rise. Commendable studies have already been made by Menard, Shumway, and Chase. Menard (1964) suggested that the northeastern trend of the Cocos Ridge, and of the Tehuantepec and Nasca submarine ridges (Fig. 1), might be related to differential movement along east-west fracture zones on the ocean floor. More recently (1966), he has suggested that the eastward bulge of the crest of the East Pacific Rise in the region west of the Galápagos Islands might be the result of right-lateral offset along the east-west Galápagos Fracture Zone. Whether the differential eastward movement of the ocean floor is caused by gravity sliding on the flanks of the East Pacific Rise or by the drag of convection currents in the mantle is still in doubt (Menard and Chase, 1965).

Many geologists have suggested that major volcanoes, particularly oceanic ones, tend to be spaced at fairly regular intervals and that these intervals are roughly equal to the thickness of the crust in each region. Lowthian Green (1887), for instance, thought that the major volcanoes of the Hawaiian Islands are spaced approximately 20 miles or multiples of 20 miles apart, and Shand (1937) thought that a similar spacing could be seen in the Galápagos and Society islands and elsewhere. Friedlaender (1918), while supporting the idea of a regular spacing, maintained that the major volcanoes of the Hawaiian, Society, and Galápagos islands lie approximately 21 to 24 miles apart, or multiples thereof. Reference to the map (Fig. 1) tends to confirm Friedlaender's calculations as applied to the Galápagos region, more particularly with respect to the spacing of volcanoes aligned in a northwest or north-northwest direction. But the cause of the spacing, which greatly exceeds the average thickness of the crust under the Pacific Ocean, remains to be determined.

VOLCANOGENIC PATTERNS

Probably no volcanoes in the world display more spectacular examples of circumferential fractures around their summit calderas or radial fractures on their flanks than do the five principal volcanoes of Albemarle Island and the mighty shield of Narborough Island. Arcuate fractures result both from tumescence and from subsidence in the wake of receding magma; radial fractures originate when volcanoes swell under the influence of rising magma.

One looks in vain for traces of radial fracture patterns on all the other islands of the archipelago, and only on Bindloe Island can one detect a crudely arcuate arrangement of some of the parasitic cones. On Charles Island, all but a few of the late cones appear to be scattered at random; on Chatham Island, the major volcano is elongated along a tectonic trend, but the smaller, younger volcanoes in the eastern part of the island reveal no systematic grouping. On James Island, the major volcano and almost all of the minor ones are arranged along northwest-trending fractures; on Abingdon Island, the cones lie mainly along fractures that trend north-northwest. Duncan Island repeats the same trend, but on Jervis Island the cones and domes show no apparent pattern. Arcuate and radial fractures appear to characterize the mature or "Mauna Loa stage" of rapid growth of the Galápagos volcanoes and are therefore virtually confined to the Albemarle and Narborough volcanoes. During the later, "Mauna Kea," stage, while the summit calderas are being buried, the minor cones show either random arrangement, or, exceptionally, as on Bindloe Island, some of them reflect an arcuate pattern inherited from the preceding stage.

We note in conclusion that the western part of Abingdon Island and of the Cape Berkeley volcano on Albemarle Island have been downfaulted below sea level. The partial collapse of the old Abingdon volcano was probably tectonic in nature, but, as Figure 17 suggests, the partial collapse of the Cape Berkeley volcano was probably caused either by submarine landslides or by engulfment along arcuate fractures when magma migrated underground.

Uplift and Subsidence

Darwin was the first to report the presence of marine fossils in the Galápagos Islands, and he thought, quite properly, that shells embedded in the tuffs on Chatham Island must have been uplifted *en masse* along with the enclosing beds and were not blown to their present position by submarine eruptions. Wolf (1895) also discovered shell-bearing beds on a few of the islands, some of them as much as 300 feet above the sea. Ochsner, while on the California Academy of Sciences Expedition of 1906, found a supposedly Pleistocene beach deposit of white shell sand near Vilamil, on Albemarle Island, about 40 feet above the sea, and also supposedly Pliocene shell-bearing limestones between uplifted submarine lavas on Indefatigable and Baltra islands. Finally, Eibl-Eibesfeldt (1959) reported the occurrence of "excavations characteristics of sea urchins" on terraces near Tagus Cove, Albemarle Island, more than 325 feet above the sea. We could not find these markings nor any other possible signs of recent uplift in this vicinity.

Our studies have revealed that upheaved submarine lavas, some of them interbedded with fossiliferous limestones, are present on Hood and Barrington islands, along the northeastern coast of Indefatigable Island, and on Baltra Island, in other words, in a narrow belt that trends approximately N.30°W. through the southern part of the archipelago. These uplifted areas, as noted already, are crossed by approximately east-west faults. When the submarine lavas were raised above sea level may never be known, but it may have been long after they were erupted. Certainly this regional uplift must have taken place less than about 1.5 m.y. ago, which is the approximate age of some of the youngest submarine lavas on Baltra Island, and it is equally clear that many of the east-west fault scarps were either produced or rejuvenated in very recent times.

Local uplifts are common in regions of recent volcanism, and it can be assumed with confidence that many such uplifts accompanied the growth of the Galápagos Islands. A spectacular instance was

101

provided as recently as 1954 at Urvina Bay, on the western coast of Albemarle Island, at the foot of Alcedo volcano, when a limestone reef rich in corals, extending about 4 miles along shore, was rapidly uplifted to a maximum height of about 15 feet, driving the coast line outward to a maximum distance of 0.75 mile (Richards, 1962). Uplift was almost certainly caused by magma rising from below, for in November 1964, about six months after the uplift, an eruption took place on the flank of the adjacent Alcedo volcano.

No convincing proof has been presented, however, for broad, regional uplifts of the archipelago as a whole, such as Dall and Ochsner (1928) proposed. In their opinion, the Panama region was uplifted at the close of the Oligocene or at the beginning of the Miocene Epoch, and "the Galápagos group or its pre-existing nucleus underwent enlargement and uplift" at that time, a process which, they thought, continued intermittently into Pleistocene time. This, however, was no more than a suggestion.

Some biologists have postulated uplifts to account for the distribution of certain kinds of animals and plants in the archipelago. It has been supposed, for example, that the five large volcanoes of Albemarle Island once formed separate islands that were subsequently linked by uplift. Admittedly, these volcanoes must have been detached from each other during their early stages of growth, but obviously as they grew they must have coalesced; the postulated uplifts are quite unnecessary. Robinson (1902) concluded from his study of plant distribution in the archipelago that the individual islands developed separately by emergence.

Signs of major uplifts, such as high marine terraces and deeply incised canyons, are completely lacking from the archipelago. This is not surprising in view of the recency of most of the surficial lavas on the islands and the fact that valleys cut by stream erosion are and must always have been both rare and small.

Some writers have maintained, for purely biological reasons, that the present archipelago was once a single island; if that is so, there must have been a submergence, at least in some parts, of about 1000 fathoms. Alban Steward (1911) was one who thought that the botanical evidence favored the idea of wholesale subsidence. Kroeber (1916), however, from a statistical analysis of the flora listed by Robinson, concluded that "the origin of the Galápagos Islands is scarcely a soluble botanical problem"; for him, the botanical evidence for submergence rather than emergence was inconclusive.

There is no geological evidence for a widespread, profound sinking, nor indeed any unequivocal proof of even a slight sinking of the islands as a whole. Parts of some of the islands, as already noted, have been dropped by faulting; for instance, the western part of the Cape Berkeley volcano and the southern part of Hood Island. But these are local subsidences, caused in part by subterranean migration of magma. One looks in vain for deeply embayed coast lines and large bay-mouth deltas, even on the most eroded islands, and nowhere are there any partly submerged coastal cliffs. If such features were present, moreover, they would not prove wholesale subsidence; they might have developed during the Pleistocene Epoch when sea level sank and then have been drowned when sea level rose. Chubb noted the presence of a partly submerged peat bog in Conway Bay, on the northwestern coast of Indefatigable Island, and thought that it might denote local, recent subsidence; but if there had been a more general, regional subsidence, the lower, gentle slopes of all the major volcanoes of the archipelago would have been flooded to produce wide, submerged shelves; these, however, seem to be absent. We conclude, therefore, as Darwin concluded, that there is no geological evidence for either regional uplift or subsidence affecting the archipelago as a whole.

Age of the Islands

Debate will long continue as to how much isolation has been necessary to account for the present character of the fauna and flora of the Galápagos Islands. Views concerning rates of evolution vary widely; some biologists demand long stretches of time to explain the present aspect of the Galápagos biota; others maintain that isolation itself, particularly in an arid climate, has greatly accelerated the evolution of new forms.

Biological evidence as to the age of the islands is vague, doubtful, and conflicting. Vinton (1951), among others, supposed that during early Tertiary times what is now the submarine Cocos Ridge was a long peninsula reaching most of the way from Costa Rica to the Galápagos region, permitting early colonization of the archipelago. Kuschel (1963), on the other hand, after comparing the terrestrial faunas on Easter, Juan Fernandez, Desventuradas, and the Galápagos islands, came to the following conclusion: "If the origin of life in the Galápagos is placed as far back as the early Tertiary, then it seems impossible to understand the much lower degree of speciation and specialization of the terrestrial organisms in comparison with those on Juan Fernandez, especially as the more favorable climate produces a higher turnover of genetic material. One would therefore expect to find an even greater difference in species between the Galápagos and the mainland than between Juan Fernandez and the continent." Most of the fauna of the Juan Fernandez and Desventuradas, in his opinion, dates back to the Eocene and part of the Oligocene period, "while the Galápagos fauna, including the terrestrial vertebrates, might go back only to the Pliocene, or, even to the end of the Pliocene and to the Pleistocene."

Dall (Dall and Ochsner, 1928) thought that the marine fossils that Ochsner collected from the limestones interbedded with the lavas on Baltra and Indefatigable islands were of Pliocene age. Professor Wyatt Durham (1964) has suggested, however, that marine fossils from older sedimentary rocks near the base of the coastal section on Indefatigable Island, close to Cerro Colorado, may be of late Miocene age. If further studies confirm this view, it seems likely that a low island was already present in this vicinity at this early date, for, as we have noted before (p. 15), the fossiliferous marine limestones near Cerro Colorado are closely associated with tuffaceous sediments that appear to have been baked and reddened by a lava flow that poured overland and with cross-bedded calcareous sandstone suggestive of deposition on land. But if such an early island was once present here, it must soon have been submerged, for the beds that may have been laid down subaerially are covered by a thick series of submarine lavas and interbedded limestones.

Valuable paleomagnetic data and potassium-argon age determinations have recently been presented by Cox and Dalrymple (1966). The two oldest lavas dated by Dalrymple are uplifted submarine flows that, according to Cox, show reversed polarity. One of the two lavas, collected at the northern end of the north-dipping section on the northeastern coast of Indefatigable Island, is 1.47 ± 0.40 m.y. old; the other, from higher in the sequence, collected near the main boat landing on Baltra Island, is 1.37 ± 0.16 m.y. old. It should be noted, however, that the first of these two lavas is closely associated with a flow showing normal polarity, the age of which is 0.74 ± 0.22 m.y. Three samples of submarine lavas from near Punta Suarez on Hood Island show reversed polarity and are therefore more than 0.7 ± 0.05 m.y. old. The big problem remains: when were these submarine lavas on Hood, Indefatigable, and Baltra islands uplifted above sea level to form an ancestral landmass? All we can say at present is that the uplift must have taken place less than 1.37 ± 0.16 m.y. ago, for this is the age of the youngest of the lavas that were certainly laid down beneath the sea. Doubt remains as to whether the lava from Indefatigable Island that has been dated as 0.74 ± 0.22 m.y. old was laid down on land or on the sea floor.

Other lavas in the archipelago that must be more than 0.7 ± 0.05 m.y. old, on the basis of their reversed polarity, are the following: two samples from Wenman and two from Duncan Island; four samples from the coastal cliffs near Daylight Point, at the northwestern corner of Charles Island; and one sample, associated with another showing normal polarity, from the northern coast of Jervis Island. We think that all of these lavas, with the possible exception of those from the northwestern corner of Charles Island, were laid down on land. All other lavas in the archipelago that were tested by Cox show normal polarity, and all other potassium-argon age determinations made by Dalrymple range only up to about 0.30 m.y. (Fig. 25).

When the Galápagos volcanoes began to grow from the ocean floor, in what order they grew, and when each first rose above sea level may never be known. All that can be said with reasonable certainty is that the oldest of the visible lavas that erupted on land and never submerged do not date back more than about 1 m.y., and most of them were erupted less than half as long ago.

Tuzo Wilson (1963) has suggested that each of the Galápagos Islands originated on or close to the crest of the East Pacific Rise and has since been drifting southeastward; hence, in his opinion, the islands become progressively older in that direction. Banfield and his

Figure 25. Paleomagnetic data from the Galápagos Islands. Shaded area contains all lavas older than the Brunhes normal polarity epoch (more than 0.85 ± 0.15 m.y.), except for those on Wolf Island, which lies northwest of the map area. Solid circles are flows of the Brunhes epoch; open circles are older flows. Adjacent flows of the same polarity are shown with one symbol. (*From* Cox and Dalrymple, 1966).

colleagues (1956) also supposed that the eastern islands are the oldest but that the archipelago is expanding westward by the addition of new volcanoes in that direction. Neither of these views is supported by our observations. Admittedly, the most copious and numerous eruptions during Holocene times have been on the western islands of Narborough and Albemarle, but there have also been countless eruptions during Holocene times on the eastern islands of Abingdon, Bindloe, and Chatham, and the easternmost island of the archipelago, namely Tower, is almost completely mantled by extremely youthful flows. However, the lavas of the westernmost islands are tholeiitic basalts typical of the mature or "Mauna Loa" stage of evolution, whereas those of the eastern islands are chiefly alkali-olivine basalts typical of a later "Mauna Kea" stage.

Possible Continental Connections

Many biologists have favored the idea of former land connections between the Galápagos Islands and either Central or South America to account for the present aspect of the fauna and flora. It seems proper, therefore, to discuss the geological evidence.

A long submarine ridge, the Cocos Ridge, stretches southwestward from Costa Rica almost to the Galápagos archipelago (Fig. 1). It has been suggested that parts of this ridge may once have formed a chain of island stepping stones leading to the archipelago, and some biologists have imagined that the entire Cocos Ridge was once a peninsula linked to Costa Rica, providing a ready means of terrestrial and coastal migration for animals and plants. Vinton (1951) thought that such a peninsula was already present during Oligocene and Miocene times, stretching to within 100 miles of the archipelago, which was then, in his opinion, a single, large landmass. At the same time, so he thought, the Isthmus of Panama was submerged, permitting currents from the Caribbean to raft organisms southwestward along the shores of the peninsula to the Galápagos. The peninsula was then submerged and not until Pleistocene times, according to Vinton, did eruptions build Cocos Island. This sequence of events seemed to him to explain most satisfactorily the differences between the faunas and floras of the Galápagos and Cocos islands.

The bathymetric map (Fig. 1) shows that the Cocos Ridge is essentially a broad, irregular rise, about 100 miles wide, most of it at depths of 800 to 1000 fathoms, but with scattered peaks rising to within 100 to 500 fathoms of sea level. On its northwestern side, the broad rise is flanked by a narrower, discontinuous ridge which Shumway and Chase (1961) regard as possibly "a zone of particular crustal instability where repeated volcanic activity" has taken place. Cocos Island is the only volcano that rises above sea level, but several submarine cones are also aligned along this discontinuous ridge. The lavas of Cocos Island are of "oceanic type," and, in common with

109

those of the Galápagos Islands, they suggest the absence of an underlying, sialic, continental crust.

No flat-topped cones (guyots), such as might suggest the submergence of former volcanic islands, have been found along the Cocos Ridge. It is not impossible, however, as Professor Menard has informed us, that the broad part of the ridge, southeast of the narrow, discontinuous volcanic ridge, may once have been a landmass that was first truncated by wave action and then subsided. But until more data become available, particularly samples collected by dredging and coring on the crest of the Cocos Ridge, it is unwise to speculate further.

Banfield and his colleagues (1956) found no proof of former continental connections nor any strong evidence favoring the idea of a once-continuous Galápagos landmass later dismembered by submergence. They did suggest, however, that volcanism in the Galápagos archipelago "moved successively southwestward while erosion planed off the eastern islands as vulcanicity declined there." With this idea in mind, they concluded that "seeding" of the islands with land life "might be explained by the growth and later marine erosion of a succession of stepping stones, developing westward, their eastern bridgelike members having later been largely destroyed." Tuzo Wilson (1963) and Adrian Richards (1966) also maintain that the Galápagos Islands become older toward the east, the islands having reached their present positions by drifting away from the East Pacific Rise. The oldest rocks yet identified are the uplifted submarine lavas in the south-central part of the archipelago, and, while it is true that the most active volcanoes of the archipelago are the westernmost, namely those on Albemarle and Narborough islands, eruptions have continued down to Holocene times on all of the eastern islands. The bathymetric map (Pl. 1) shows several submerged cones east and northeast of the archipelago, but they are widely spaced and there is no proof that any of them were formerly islands.

The Galápagos volcanoes rise from the western end of a broad submarine ridge that extends almost to the mainland of Ecuador, separated only by a narrow coastal shelf and by the narrow and deep Peruvian Trench. The eastern part of this ridge has been named the Carnegie Ridge and the western part the Galápagos Platform. These parts are separated by a saddle, the lowest points of which lie at depths of between 1200 and 1300 fathoms (Shumway and Chase, 1961). Most of the Carnegie Ridge and Galápagos Platform lies at depths of less than 1000 fathoms, and locally there are conical peaks, presumably submarine volcanoes. Three of these peaks lie 45 to 90 miles east of the easternmost of the Galápagos Islands, but there is no evidence that any of them ever rose above sea level nor that any other parts of the Carnegie Ridge and Galápagos

Platform ever did so. Here again, as in the case of the Cocos Ridge, there is urgent need for samples collected by coring and dredging.

Nygren (1950) suggested that, from Eocene through early Miocene times, the axis of greatest accumulation of marine sediments in Ecuador lay on the present site of the coastal lowlands, and he surmised from the nature of the sediments that the basin of accumulation was flanked on the west, that is, on the present site of the easternmost part of the Carnegie Ridge, by a wide, low landmass. He supposed that this landmass foundered beneath the sea during the orogeny that began during middle Miocene time, when the eastern borderland rose to form the present cordilleras of Ecuador. It seems probable, however, that even during early Tertiary times the extensions of continental lands to the west of Ecuador and Colombia were minor, and, even supposing that the Galápagos Islands were then in existence, they would still have been isolated by at least 400 miles of ocean.

Floating Pumice as a Means
of Biological Dispersal

At least two Galápagos volcanoes have erupted trachyte pumice, namely Duncan Island and Alcedo volcano on Albemarle Island; of these, the more recent and more voluminous eruption was that of Alcedo volcano. Waterworn lumps of pumice are plentiful on many of the beaches of the archipelago; they are present, for example, on the beaches of James, Indefatigable, Bindloe, and Tower islands. The samples we examined from the first three of these islands are hedenbergite trachytes, essentially identical to the pumice erupted by Alcedo volcano; the pumice from Tower Island was not examined. Despite these observations, we feel reasonably sure that large quantities of floating pumice must have reached the Galápagos Islands from the continent, carried there by the equatorial currents. Innumerable eruptions of pumice have taken place during and since late Tertiary time from the volcanoes of Mexico and Central and South America, and it can hardly be doubted that much of this material was drifted westward to the archipelago. Future petrographic studies may confirm this view by revealing waterworn lumps of rhyolitic or dacitic pumice, typical of the continental volcanoes, or of pumice characteristic of the Revillagigedo volcanoes off the coast of Mexico. But even if such lumps can no longer be found, having been destroyed by beach erosion, their former presence is virtually certain. We think, therefore that consideration should be given to the idea of floating pumice as one of the means whereby some animals and plants reached the Galápagos Islands.

Darwin and others, members of the "flotsam and jetsam" school, thought that many organisms reached the archipelago on floating logs or on mats of uprooted vegetation. H. W. Bates (1892), one of the earliest explorers of the Amazon, seems to have been the first to suggest floating pumice as a means of dispersal, and we are grateful to our colleague, Professor Clarence Palmer, of the Institute of

Geophysics, University of California, Los Angeles, for drawing our attention to Bates' comments on this problem. Natives had told Bates that lumps of pumice, "frozen foam" as they called them, could be seen floating down the Amazon, more than 2000 miles from their source in the Ecuadorian Andes, and he felt sure that many of these lumps were carried down to the Atlantic Ocean, to be dispersed by currents to distant shores. "The probability of these porous fragments serving as vehicles for the transportation of seeds of plants, eggs of insects, spawn of fresh-water fish, and so forth, has suggested itself to me. . . . The eggs and seeds of land insects and plants might be accidentally introduced, and safely enclosed with particles of earth in the cavities." After returning to England, he wistfully confessed that he had failed to find out whether or not such transport actually takes place.

For every log and for every raft of matted vegetation floated to the Galápagos Islands, it seems reasonably certain that tens of thousands of lumps of pumice must have drifted there within the last few million years. No large rivers, like the Amazon and Orinoco, traverse heavily forested regions to reach the western shores of Central and South America, although many smaller ones cross wooded highlands to reach the shores of southern Costa Rica, Panama, and Colombia. Vast quantities of pumice have been carried by rivers to the Pacific Ocean from the volcanoes of Mexico, Central America, and the Andes, and much has been washed into the ocean from the beaches. Some of this pumice, having lain for a time on river banks and beaches, must have accumulated soil and vegetation, along with seeds, eggs, and the young of various animals and plants. Once this populated pumice reached the open ocean, it may have drifted to the Galápagos Islands within a few months.

The trans-Pacific distribution of floating pumice from the 1952 eruption of Barcena volcano in the Revillagigedo Islands (110°W., 20°N.) has been studied by Richards (1958). Abundant lumps reached the beaches of Hawaii after 264 days, with a mean drift rate of 22 cm/sec. Some pumice reached Johnston Island, 6100 km. from Barcena, in 225 days, including several hundred pounds of rounded pieces up to 7.5 cm in diameter; some reached Wake Island and the Marshall Islands in 560 days, having traveled 8700 km. One lump that landed on Wake Island measured 19 cm in length, although most are said to have been of walnut and potato size. Dense formations of floating pumice were seen in the open sea close to the western Caroline Islands, all bearing a growth of barnacles, and many lumps that reached the Marshall Islands had small corals attached to them.

Marine dispersal of pumice, supposedly from the March 1962 eruption in the South Sandwich Islands, was studied by Sutherland (1965). The first reported strandings on the western coast of Tasmania were in late December 1963 or early January 1964. During

January 1965, pumice gravel was washed up on the southern coast of western Australia. One "pumice raft" traveled 8000 miles at an average rate of 18 miles/day. The larger lumps, being more responsive to wind assistance, traveled faster than the smaller ones, but the average drift of even fine "pumice gravel" was between 6 and 7 miles/day.

The foregoing information concerning rates of dispersal suggests that lumps of pumice might reach the Galápagos beaches from the coasts of Central and South America or from the Revillagigedo Islands in 3 or 4 months; almost surely they would arrive in less than 6 months. We need to know more, however, not only about the present ocean currents leading to the Galápagos Islands but about those during Pliocene-Pleistocene times. What kinds of animals and plants might land in the archipelago in a viable condition, as young or as seeds and eggs, we must leave for others to decide.

It may be of interest to recall in this connection that Whitaker and Carter (1954) demonstrated that the white-flowered gourd, *Lagenaria siceraria* (Mol.), generally considered to be a native of the Old World, possibly Africa, reached the Americas in pre-Columbian times, arriving in Peru as early as 3000 B.C. They showed experimentally that these gourds can float in sea water for periods up to 224 days with no significant decrease in viability of the seed, a period long enough to drift them from Africa to Brazil. They found, in addition, that up to 95 days' immersion in sea water did not impair the viability of the seeds themselves. We do not know enough about the effects of long-continued immersion in sea water on the seeds of other plants, but we hazard the guess that some kinds of plants may well have reached the Galápagos Islands on lumps of floating pumice.

One of our botanical colleagues on the Galápagos expedition, Professor Ira Wiggins of Stanford University, in answer to our queries, replied as follows (January 12, 1965) :

The great buoyancy of a large mass of pumice could well provide an excellent raft and at the same time provide good protection for seeds and drought-resistant vegetative propagules, for these parts of many plants can withstand long periods of drought if they actually remain physically dry. The pumice could insulate the seeds and other parts from overexposure to the tropical sunlight. . . . I agree with you that the pumice could easily have been a most effective agent of dispersal westward, and it probably was, along with floating mats of vegetation, hollow logs, and high level wind storms of high intensity. . . . It would require only one gravid female, or a seed per century to bring about the population of a barren island.

But neither Professor Wiggins nor Professor R. C. Stebbins, herpetologist at the University of California, Berkeley, another member of the expedition, favors the idea of transporting terrestrial vertebrates to the Galápagos Islands by means of floating pumice. Professor Stebbins thinks that the tortoises may have arrived with no more aid than that supplied by ocean currents. Iguanas, lava lizards, and snakes could survive long enough without food to make the journey, but Professor Stebbins thinks that they would almost certainly jump off the floating pumice and so be drowned. Transport by floating logs and mats of vegetation seems to him a much more likely means of dispersal of such animals. Frogs, toads, and other amphibians, which are absent from the Galápagos Islands, would also be carried in this way, and many probably reached the archipelago only to die from lack of a proper ecological niche. Professor Wiggins adds that "one factor mitigating against the transport of amphibians across 500 miles or so of salt water, either on logs or pumice, is the slight chance that such creatures would have to survive even intermittent wetting with salt water. I doubt that their systems would be able to withstand the unfavorable displacement of water from their tissues when they came in contact with the higher osmotic pressure of the salt water. They would simply be dessicated after short exposure, dry out physiologically, and die." In Professor Wiggins' opinion, amphibians are more likely to have reached the Galápagos Islands aboard rafts of vegetation, particularly if they "were to live in the basal cups of a large orchid plant or in the spathe of an aroid."

Our conclusion is that floating pumice is not likely to have carried terrestrial vertebrates to the Galápagos Islands, but it may well have carried many kinds of plants and insects. The problem appears to us to be one that deserves to be tested by experiment and observation. However, arrival in the islands is one problem; survival is another, and often more difficult, problem!

Petrology

Some petrographic and chemical features of the Galápagos lavas have already been pointed out in our descriptions of individual islands. Here we group the rocks according to their salient characteristics and petrologic relations. We consider first the oldest lavas exposed on the uplifted platform underlying the south-central islands of the archipelago, then turn to the younger basalts, and finally to the differentiated lavas and plutonic rocks.

UPLIFTED SUBMARINE LAVAS

Uplift of a southeast-trending belt extending from Baltra Island along the eastern coast of Indefatigable, through Barrington, to Hood Island has exposed shallow submarine lavas, some of them interlayered with fossiliferous sedimentary rocks. These lavas are the oldest in the archipelago and probably represent the upper par of the Galápagos Platform. Wherever exposed in shoreline cliffs, they are nearly flat-lying sheets, averaging 3 or 4 m in thickness. Pillow structures are extremely rare, but the presence of fossiliferous interbeds leaves no doubt that the flows were poured out beneath the sea.

All of these uplifted lavas are basalts, but they exhibit moderate petrographic and chemical diversity. Among them are tholeiitic, alkaline, and possibly high-alumina basalts, the first two types differing little from late-stage Holocene basalts throughout the archipelago. Chemical analyses of representative specimens are presented in Table 1.

Typical of the tholeiitic basalts are those of Baltra Island, described on page 20. These are aphyric lavas containing intermediate plagioclase and subcalcic augite; olivine is rare and most of it is rimmed with clinopyroxene. Two analyzed specimens (Nos. 28, 29) contain normative quartz. These lavas are distinctly less basic than the younger Galápagos basalts. They contain plagioclase zoned to andesine, as well as interstitial anorthoclase, and their normative plagioclase in andesine. They are also among the most iron-rich

117

TABLE 1. CHEMICAL ANALYSES OF UPLIFTED SUBMARINE LAVAS
OF BALTRA, HOOD, AND BARRINGTON ISLANDS

	28	29	31	102	111
SiO₂	48.26	48.13	46.21	45.99	46.28
TiO₂	3.32	3.53	1.77	1.73	1.42
Al₂O₃	14.07	13.29	16.22	16.62	16.46
Fe₂O₃	4.22	4.72	2.11	2.20	3.00
FeO	9.43	9.53	8.16	7.53	7.97
MnO	0.15	0.22	0.16	0.17	0.29
MgO	5.69	5.63	9.18	10.75	9.62
CaO	10.64	10.11	10.99	10.50	11.41
Na₂O	2.88	3.10	2.38	2.59	2.29
K₂O	0.59	0.63	0.30	0.45	0.13
H₂O+	0.70	0.60	0.95	0.76	0.71
H₂O—	0.14	0.13	1.22	0.21	0.10
P₂O₅	0.24	0.43	0.18	0.21	0.09
Other	..	..	..	..	..
Total	100.33	100.05	99.83	99.71	99.77
Ap	0.46	0.94	0.25	0.40	0.18
Il	4.74	5.06	2.52	2.40	2.00
Or	3.60	3.85	1.80	2.65	0.80
Ab	26.40	28.65	21.85	21.20	20.70
An	24.20	21.13	33.33	32.20	34.43
C	..	..	..	..	..
Mt	4.51	5.07	2.10	2.59	3.15
Hm	..	..	..	..	..
Di	22.88	22.12	17.36	14.84	17.68
Hy	12.22	12.18	3.52	..	4.00
Ol	..	..	17.22	22.79	17.10
Q	1.05	1.00	..	..	..
Ne	..	..	..	1.23	..
Total	100.06	100.00	100.01	100.00	100.04
si	119.1	114.2	101.8	97.6	99.3
al	20.2	18.5	21.1	20.7	20.7
fm	44.2	47.7	47.3	49.6	48.1
c	27.9	25.7	26.1	23.8	26.2
alk	7.7	8.1	5.5	5.9	5.0
k	0.120	0.118	0.077	0.103	0.037
mg	0.472	0.421	0.642	0.692	0.643
qz	—11.7	—18.2	—20.2	—26.0	—27.0

28. Olivine-augite basalt, lowermost lava in cliffs near pier on western side
of Baltra Island.
29. Analcite-bearing olivine basalt, directly overlying No. 28.
31. Analcite-bearing alkali-olivine basalt higher in same section as 28 and 29.
102. Alkali-olivine basalt from peak 2.5 miles east of Punta Suarez, Hood
Island.
111. Olivine-bearing high-alumina basalt, northeastern corner of Barrington
Island.

rocks we collected. In these respects, they approach hawaiites (Macdonald, 1960), but, because they are somewhat less rich in alkalis, we prefer to call them ferrobasalts.

The lavas of Barrington Island, as noted earlier (p. 14), share petrographic and chemical features with high-alumina basalts, as defined by Kuno (1960). An analyzed specimen (No. 111, Table 1) falls within the "wishbone" field of high-alumina basalts of appropriate silica content (Kuno, 1960, Fig. 10). Few high-alumina basalts are as poor in silica as are these Barrington Island lavas, although the Modoc basalts of northern California and southern Oregon are only slightly more siliceous.

A distinctive feature of the Barrington Island basalts is their oikocrysts of clinopyroxene, illustrated in Figure 3a. Similar textures are to be seen in lava fragments among the explosive ejecta of cones on James Island, but the composition of one such fragment, from the large cone bordering Sullivan Bay (No. 24, Table 2c) is distinctly alkaline and contains less alumina than does the analyzed Barrington basalt.

Hood is the southernmost island where submarine lavas have been exposed by uplift. The basalts here are strongly alkaline, containing normative nepheline, modal analcite and other zeolites, and abundant olivine (see p. 11 and Fig. 3c). An analysis of the illustrated specimen (No. 102) is given in Table 1.

The youngest of the uplifted lavas on Baltra Island are transitional between the slightly older tholeiitic ferrobasalts of that island and the strongly alkaline basalts of Hood Island. Specimen No. 31 (Table 1) is such a lava, collected from near the top of the uplifted series on the western side of Baltra Island. Its mineral assemblage is typical of alkali-olivine basalts, namely olivine, with inclusions of brown spinel, ophitic titaniferous augite, sodic labradorite, analcite, ilmenite, and magnetite. Nevertheless, it contains a small amount of normative hypersthene that cannot be accounted for by oxidation of iron, and it is notably poorer in alkalis than are the underlying tholeiites represented by Nos. 28 and 29 in the same table.

YOUNGER BASALTS

The younger Galápagos basalts, those that form the emergent volcanoes of the archipelago, can be separated into three principal groups, namely tholeiites, olivine tholeiites, and alkali-olivine basalts. Each group has distinctive chemical and petrographic characters and each predominates in one of three groups of islands. Our classification of these basalts is based on their normative compositions, primarily on the presence or absence of normative nepheline, olivine, or hypersthene. This classification, despite certain inadequacies, provides the most convenient means of emphasizing important differences between the groups, without introducing a new nomenclature.

Tholeiites

Tholeiitic basalts make up most, although not all, of the huge shield volcanoes of the western islands, Narborough and Albemarle.

They are also found on James and Jervis islands. Most of them contain sparse phenocrysts of labradorite, but only a few contain phenocrysts of olivine. Their characteristic clinopyroxene is augite, but pigeonite is present in the groundmass of a few specimens. All contain normative hypersthene and most contain normative quartz. Descriptions of typical examples have already been given in our discussions of Narborough and Albemarle, and analyses of representative specimens are given in Table 2a. Previously published analyses are given in Table 3, an average of four analyses is given in Table 4, and trace elements of one specimen (No. 42) are listed in Table 7.

High titania seems to be characteristic of all modern analyses of Galápagos tholeiites and is especially pronounced in iron-rich rocks of Narborough, Albemarle, and James islands. The low values reported by Richardson (1933), reproduced in Table 3, are presumably due to less accurate determinations of TiO_2 in analyses made before the spectrophotometer came into general use. TiO_2 varies inversely with MgO and directly with total iron up to a point of maximum iron enrichment, as shown in Figure 26.

Olivine Tholeiites

The tholeiites of the eastern islands are somewhat higher in magnesium and have lower iron-magnesium ratios than those of the western islands. They are also poorer in Ti, Sr, Zr, Cu, and V than the typical tholeiite of the western islands. They contain normative olivine rather than quartz. Analyses of three specimens are given in Table 2b (Nos. 10, 15, 35) and trace elements of one of these (No. 35) are given in Table 7. An average of three rocks is given in Table 4.

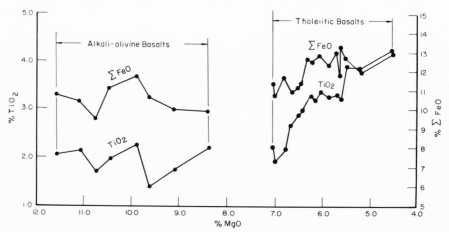

Figure 26. Relation of TiO_2 and total iron (as FeO) to MgO content of alkali-olivine basalts and tholeiites.

TABLE 2. CHEMICAL ANALYSES OF RECENT BASALTS

a. Western Islands

	42	56	63	87	132	134
SiO$_2$	48.79	47.20	48.45	48.46	48.35	47.84
TiO$_2$	3.02	3.35	3.39	3.05	3.21	3.19
Al$_2$O$_3$	14.72	15.24	13.75	14.96	13.76	13.42
Fe$_2$O$_3$	2.80	3.50	4.72	2.59	2.58	6.97
FeO	8.91	8.42	8.60	9.23	10.39	6.14
MnO	0.16	0.17	0.20	0.21	0.11	0.10
MgO	6.43	6.37	6.05	6.38	6.27	6.21
CaO	11.24	10.45	10.71	11.58	10.59	11.56
Na$_2$O	2.62	3.11	2.79	2.26	2.70	2.63
K$_2$O	0.42	0.70	0.50	0.38	0.46	0.38
H$_2$O+	0.39	0.78	0.36	0.55	0.73	1.17
H$_2$O—	0.01	0.11	0.02	0.05	0.22	0.39
P$_2$O$_5$	0.31	0.40	0.36	0.29	0.26	0.21
Other	..	..	..	..	..	..
Total	99.82	99.80	99.90	99.99	99.63	100.21
Ap	0.66	0.80	0.72	0.60	0.56	0.46
Il	4.28	4.76	4.84	4.34	4.62	4.60
Or	2.55	4.25	3.05	2.30	2.80	2.35
Ab	23.95	28.60	25.65	20.70	25.05	24.50
An	27.75	26.15	24.08	30.25	24.83	24.52
C	..	..	..			..
Mt	2.99	3.74	5.05	2.77	2.78	7.54
Hm						
Di	21.84	19.68	22.80	21.44	22.08	26.80
Hy	15.32	4.54	11.68	16.24	17.28	4.92
Ol	..	7.56	..	..	..	..
Q	0.64	..	2.11	1.36	..	4.31
Ne	..	..	..	..	..	..
Total	100.00	100.08	99.98	100.00	100.00	100.00
si	114.0	113.5	119.0	115.3	114.0	112.1
al	20.3	21.5	19.8	20.9	19.1	18.5
fm	45.0	43.3	44.7	43.8	47.3	46.1
c	21.8	26.9	28.2	29.5	26.7	28.9
alk	6.6	8.3	7.4	5.8	6.9	6.5
k	0.096	0.134	0.150	0.100	0.101	0.087
mg	0.500	0.530	0.500	0.522	0.468	0.473
qz	—12.4	—19.7	—10.6	—2.9	—13.6	—13.9

42. Tholeiitic basalt from aa flow at an elevation of 1800 feet on the southwestern slope of Narborough volcano.

56. Olivine basalt, large aa flow at Cape Berkeley, Albemarle Island.

63. Tholeiitic basalt, block from Tagus Cone, Albemarle Island. Host rock of eucrite inclusion, No. 64, Table 5.

87. Olivine-augite basalt, pahoehoe flow at Punta Espinoza, Narborough Island.

132. Tholeiitic basalt from aa flow at Alcedo volcano, eastern side of Ismo Parry, Albemarle Island.

134. Tholeiitic basalt underlying pumice of Alcedo volcano near eastern coast of Albemarle Island.

TABLE 2. CHEMICAL ANALYSES OF RECENT BASALTS (Continued)

b. Northeastern Islands

	4	5	5g	8	10	15	35
SiO_2	48.10	46.62	47.52	48.49	48.22	48.09	48.68
TiO_2	2.88	1.01	2.70	2.74	2.30	2.27	1.96
Al_2O_3	14.68	26.26	14.91	15.43	13.84	15.43	15.22
Fe_2O_3	2.19	1.65	3.52	1.69	3.74	1.35	3.71
FeO	9.13	3.07	7.80	9.15	9.11	10.19	7.43
MnO	0.20	0.06	0.21	0.17	0.19	0.21	0.20
MgO	6.65	3.10	6.70	6.67	6.78	7.02	7.01
CaO	11.21	14.46	11.28	11.40	11.69	11.54	11.72
Na_2O	3.60	2.53	3.48	3.02	3.16	2.79	2.55
K_2O	0.71	0.34	0.69	0.80	0.29	0.29	0.34
H_2O+	0.23	0.59	0.75	0.17	0.52	0.32	0.66
H_2O-	0.22	0.12	0.12	0.13	0.31	0.08	0.09
P_2O_5	0.37	0.18	0.41	0.32	0.18	0.25	0.24
Other	..	..	..	..	..	..	..
Total	100.17	99.99	100.09	100.18	100.33	99.83	99.81
Ap	0.78	0.22	0.88	0.68	0.37	0.53	0.50
Il	4.04	1.40	3.80	3.84	3.26	3.20	2.78
Or	4.25	2.00	4.15	4.75	1.75	1.75	2.05
Ab	26.50	18.65	27.60	27.08	28.85	25.40	23.35
An	21.97	59.25	23.23	26.35	23.10	29.08	29.68
C	..	..	..	..	..	..	..
Mt	2.25	1.73	3.72	1.77	3.97	1.43	3.93
Hm	..	..	..	..	..	..	..
Di	25.40	9.76	24.56	22.80	27.72	21.36	22.44
Hy	..	..	..	..	1.92	5.38	15.30
Ol	11.25	4.52	9.64	12.57	9.06	11.88	..
Q	..	..	..	..	..	..	0.02
Ne	3.63	2.46	2.43	0.16	..	..	..
Total	100.07	99.99	100.01	100.00	100.00	100.01	100.05
si	117.7	112.6	111.4	110.7	108.2	109.5	114.6
al	20.0	37.3	20.6	20.8	18.2	20.7	21.1
fm	43.1	19.0	42.2	43.5	46.4	44.8	43.1
c	27.8	37.3	28.3	27.9	28.0	27.9	29.5
alk	9.1	6.4	8.9	7.8	7.4	6.6	6.3
k	0.115	0.081	0.116	0.149	0.057	0.064	0.081
mg	0.538	0.590	0.558	0.525	0.491	0.533	0.576
qz	—24.7	—13.0	—24.2	—19.5	—21.4	—16.9	—10.6

4. Low-magnesium alkali basalt, nonporphyritic portion of composite lava near southeastern end of Abingdon Island.

5. Strongly porphyritic "big-feldspar basalt" from same composite flow as 4.

5g. Groundmass of 5.

8. Low-magnesium alkali basalt from very recent pahoehoe lava at northwestern end of Abingdon Island.

10. Aphyric tholeiite, southern coast of Bindloe Island.

15. Aphyric tholeiite, recent caldera-filling lava in summit region of Bindloe Island.

35. Aphyric tholeiite, block in uppermost pyroclastic layer of Culpepper Island.

TABLE 2. CHEMICAL ANALYSES OF RECENT BASALTS (Continued)

c. Central and Southern Islands

	1	20	24	68b	76	103	110
SiO_2	46.14	47.16	45.12	47.96	45.30	46.95	45.51
TiO_2	2.01	2.28	2.07	3.29	1.73	2.22	2.14
Al_2O_3	16.10	15.15	15.48	15.03	13.99	16.58	15.06
Fe_2O_3	2.26	1.40	1.85	3.63	1.64	3.37	3.88
FeO	9.07	10.48	9.08	9.06	10.12	6.90	6.92
MnO	0.19	0.16	0.19	0.20	0.17	0.16	0.17
MgO	10.43	9.89	11.57	5.87	14.39	8.37	11.07
CaO	9.19	9.47	10.89	9.87	9.14	10.64	9.46
Na_2O	3.64	2.88	2.92	3.19	2.91	3.14	3.40
K_2O	0.28	0.79	0.18	0.68	0.24	0.74	1.14
H_2O+	0.40	0.19	0.44	0.57	0.16	0.36	0.51
H_2O-	0.08	0.01	0.10	0.11	0.06	0.27	0.10
P_2O_5	0.24	0.27	0.13	0.43	0.15	0.32	0.36
Other	..	..	..	..	..	..	..
Total	100.03	100.13	100.02	99.89	100.00	100.02	99.72
Ap	0.50	0.89	0.26	0.94	0.29	0.61	0.70
Il	2.76	3.10	2.86	4.68	2.34	3.10	2.96
Or	1.65	4.40	1.05	4.15	1.40	4.40	6.65
Ab	23.64	25.95	15.95	29.35	18.32	26.00	18.18
An	26.40	28.92	28.27	25.27	23.75	28.92	22.20
C	..	..	..	..	..	..	..
Mt	2.33	3.51	1.90	3.89	1.67	3.52	4.02
Hm							
Di	13.76	16.60	19.56	17.56	15.76	17.76	17.80
Hy					11.18		
Ol	23.82	15.31	24.05	3.06	31.62	14.34	20.15
Q				..			
Ne	5.26	1.32	5.97	..	4.33	1.35	7.21
Total	100.13	100.00	99.87	100.08	99.48	100.00	99.87
si	102.5	103.0	91.1	117.2	87.8	106.0	97.1
al	20.0	21.4	18.4	21.7	15.9	22.0	18.9
fm	51.2	45.9	52.1	43.8	59.4	44.5	51.1
c	20.9	25.0	23.6	25.9	18.9	25.8	21.6
alk	7.9	7.7	5.9	8.6	5.8	7.7	8.5
k	0.049	0.135	0.039	0.124	0.052	0.135	0.181
mg	0.646	0.598	0.674	0.492	0.703	0.638	0.693
qz	—29.1	—27.8	—32.5	—17.2	—35.4	—24.8	—36.9

1. Alkali-olivine basalt between laboratory and pier at Darwin Station, Indefatigable Island.

20. Alkali-olivine basalt, large recent pahoehoe flow at James Bay, James Island.

24. Alkali-olivine basalt with ophitic pyroxene, fragment in palagonitic cone at Sullivan Bay, James Island.

68b. Platy basalt near southern side of caldera wall, Duncan Island.

76. Picrite basalt, northern side of Buccaneer Cove, James Island.

103. Alkali-olivine basalt between Wreck Bay and Progreso, Chatham Island.

110. Alkali-olivine basalt, Rada Black Beach, Charles Island. Host lava of inclusions 110i and 98.

TABLE 3. PREVIOUSLY PUBLISHED ANALYSES OF GALÁPAGOS ROCKS

	1	2	3	4	5	6	7	8	9	10
SiO_2	61.90	46.72	47.80	48.65	45.55	48.10	48.24	47.01	47.12	48.3(
TiO_2	0.25	1.96	2.40	2.10	1.80	1.90	3.88	3.20	3.44	2.9!
Al_2O_3	16.75	14.10	18.31	17.50	18.25	15.90	15.82	15.57	14.93	13.3!
Fe_2O_3	2.27	2.14	1.47	0.25	7.28	1.93	0.78	2.32	2.57	4.7(
FeO	4.83	9.63	8.20	9.75	5.01	10.28	9.84	11.57	13.06	10.3!
MnO	0.30	0.14	0.14	0.25	0.30	0.29	0.20	0.20	0.27	0.1!
MgO	0.57	11.64	4.89	6.61	6.00	6.28	5.84	5.25	5.25	5.3(
CaO	2.30	10.92	13.00	11.85	10.20	11.60	9.84	9.77	9.44	9.6(
Na_2O	7.20	1.83	2.48	2.15	3.18	2.68	3.63	3.00	2.57	2.4(
K_2O	3.25	0.30	0.57	0.38	0.85	0.30	0.64	0.31	1.018	0.7!
H_2O+	0.20	0.20	0.31	tr	0.95	0.40	0.72	1.40	0.36	1.7!
H_2O-	0.10	0.21	0.11	0.10	0.25	0.10	0.11	0.24	0.01	0.1!
P_2O_5	0.07	0.34	0.41	0.21	0.23	0.23	0.16	0.32	0.36	0.2!
Total	100.07	100.13	100.09	99.80	99.85	99.99	99.70	100.76	100.67	100.6!
Ap	0.12	0.72	0.85	0.46	0.48	0.48	0.34	0.70	0.78	0.5!
Il	0.34	2.72	3.38	2.84	2.56	2.74	5.48	4.62	6.50	4.3(
Or	18.94	1.80	3.40	2.30	5.15	1.85	3.85	1.90	6.10	4.8(
Ab	58.83	16.40	22.55	19.55	28.60	24.85	31.95	27.85	23.55	23.1!
An	3.73	29.38	37.60	37.32	33.63	31.40	25.50	29.02	26.75	24.2!
C	..	..	..	..	..	..	..	..	..	
Mt	2.33	2.24	1.54	0.25	7.75	1.86	0.82	2.51	2.73	5.2(
Hm	..	..	..	..	..	..	..	..	..	
Di	5.72	18.08	20.20	16.60	13.24	21.16	18.60	15.24	14.92	19.2!
Hy	4.64	12.10	4.90	13.84	..	10.32	..	12.40	14.32	15.08
Ol	..	16.56	5.58	6.84	8.29	5.34	12.74	5.76	4.35	..
Q	5.35	..	..	..	..	..	..	..	..	3.37
Ne	..	..	..	..	0.30	..	0.72	..	..	
Total	100.00	100.00	100.00	100.00	100.00	100.00	100.00	100.00	100.00	100.0(
si	226.5	94.9	112.0	111.2	103.3	113.0	116.2	111.5	109.7	116.3
al	36.1	16.9	25.2	23.5	24.3	22.0	22.5	21.7	20.5	19.0
fm	21.9	55.3	36.0	42.2	42.8	42.3	42.7	46.0	48.7	49.1
c	9.0	23.8	32.4	29.0	24.7	29.2	25.5	24.9	23.5	24.9
alk	33.1	4.0	6.4	5.3	8.2	6.5	9.4	7.4	7.3	7.0
k	0.230	0.098	0.131	0.105	0.150	0.069	0.104	0.064	0.206	0.171
mg	0.143	0.642	0.478	0.538	0.477	0.440	0.495	0.407	0.376	0.392
qz	—5.9	—21.1	—13.6	0.0	—29.5	—13.0	—21.4	—18.1	—19.5	—11.7

1. Soda trachyte, James Island, collected by Darwin, analyzed by Herdsman (Richardson, 1933). Total includes 0.08% S.
2. Doleritic basalt, James Island, analyzed by Raoult (Lacroix, 1927).
3. Plagioclase basalt, Tagus Cove, Albemarle Island, analyzed by Raoult (Lacroix, 1927).
4. Porphyritic olivine basalt, Narborough Island, analyzed by Herdsman (Richardson, 1933).
5. Olivine "andesite," Chatham Island, analyzed by Herdsman (Richardson, 1933).
6. "Andesite," ejected fragment, Albemarle Island, analyzed by Herdsman (Richardson, 1933).
7. Andesine basalt, Eden Islet, analyzed by Keyes (Washington and Keyes, 1927).
8. Scoriaceous plagioclase basalt, Sierra Negra volcano, Albemarle Island, analyzed by Wiik (Banfield and others, 1956). Total includes 0.06% S, 0.46% SO_3, and 0.08% Cl.
9. Plagioclase basaltic tuff, Sierra Negra volcano, Albemarle Island, analyzed by Wiik (Banfield and others, 1956). Total includes 0.04% S, 0.15% SO_3, and 0.09% Cl.
10. Plagioclase basalt, Sierra Negra volcano, Albemarle Island, analyzed by Wiik (Banfield and others, 1956). Total includes 0.09% S, 0.24% SO_3, and 0.05% Cl.

TABLE 4. AVERAGE COMPOSITIONS OF UNDIFFERENTIATED BASALTS

	1	2	3	4	5	6
SiO_2	48.38	48.33	49.36	46.17	48.04	46.46
TiO_2	3.12	2.18	2.50	1.95	2.77	3.01
Al_2O_3	14.22	14.83	13.94	15.96	15.01	14.64
Fe_2O_3	3.71	2.93	3.03	2.23	2.71	3.27
FeO	8.67	8.91	8.53	8.26	8.69	9.11
MnO	0.26	0.20	0.16	0.19	0.19	0.14
MgO	6.32	6.94	8.44	10.11	6.67	8.19
CaO	11.24	11.65	10.30	10.32	11.30	10.33
Na_2O	2.55	2.83	2.13	2.91	3.37	2.92
K_2O	0.41	0.31	0.38	0.50	0.73	0.84
H_2O+	0.71	0.50	..	0.54	0.38	..
H_2O-	0.17	0.16	..	0.26	0.16	..
P_2O_5	0.27	0.22	0.26	0.20	0.37	0.37
Total	100.03	99.99	99.03	99.60	100.39	99.28
Ap	0.59	0.48	0.56	0.42	0.78	0.78
Il	4.46	2.86	3.54	2.72	3.88	4.24
Or	2.50	1.85	2.30	2.95	4.35	5.05
Ab	23.55	25.85	19.55	21.55	27.15	25.55
An	26.85	27.27	27.82	28.97	23.80	24.60
Mt	3.97	3.12	3.22	2.33	2.85	3.45
Di	23.04	24.12	18.16	16.88	24.28	19.92
Hy	13.66	7.40	22.94	..	..	..
Ol	..	7.05	..	21.42	10.93	15.81
Q	1.38	..	1.91	..	..	..
Ne	..	..	..	2.76	1.98	0.60
Total	100.00	100.00	100.00	100.00	100.00	100.00
si	112.7	109.2	112.8	97.2	108.8	102.1
al	19.6	19.8	18.7	19.8	20.0	18.9
fm	45.9	45.3	50.8	50.3	44.1	49.3
c	28.1	28.2	25.2	23.3	27.5	24.4
alk	6.4	6.7	5.3	6.6	8.4	7.4
k	0.106	0.068	0.105	0.101	0.125	0.160
mg	0.480	0.513	0.571	0.638	0.514	0.548
qz	—12.9	—17.6	—8.4	—29.2	—24.8	—31.7

1. Tholeiite, western type. Average of 42, 87, 132, and 134.
2. Olivine tholeiite, eastern type. Average of 10, 15, and 35.
3. Average of 181 Hawaiian tholeiites (Macdonald and Katsura, 1964, Table 9, No. 8).
4. Alkali-olivine basalt, average of 1, 20, 24, 31, 102, 103, 110, and 111.
5. Low-MgO alkali basalt, average of 4, 5g, and 8.
6. Average of 28 alkalic basalts of Hawaii (Macdonald and Katsura, 1964, Table 10, No. 4).

Although olivine tholeiites and tholeiites with normative quartz fall into two distinct groups (Fig. 29) and each type is characteristic of one group of islands, there are examples of olivine tholeiites on the western islands, and one of the tholeiites of the eastern group contains a small amount of normative quartz. On a standard AMF diagram (Fig. 27) they also tend to fall into two groups. The tholeiites, which are lower in alkalis, form a smooth curve nearly parallel to the magnesia-iron side of the diagram, while most olivine tholeiites form a second group slightly richer in alkalis with a more limited range of composition and less systematic distribution.

Alkali-Olivine Basalts

These lavas seem to be restricted to the southern and central islands and are best seen on Charles Island, where they are by far the dominant rocks. They contain abundant olivine, both as phenocrysts and in the groundmass, and many crystals include octahedra of picotite. The typical groundmass clinopyroxene is a reddish-brown to purplish-brown, titaniferous diopsidic augite. Phenocrysts of plagioclase are uncommon. Even aphyric rocks are unusually rich in magnesia and contain at least 2 or 3 percent of normative nepheline. Petrographic notes have already been presented in our accounts of Charles, Indefatigable, and James islands; analyses of representative specimens are given in Table 2c. An average of eight analyses is given in Table 4. Data on the trace elements of two samples (Nos. 1, 110) are shown in Table 7.

Alkali-olivine basalts seem to have been the only lavas erupted throughout the emergent history of the volcanoes of the southern group of islands. On James Island, however, basalts of this type

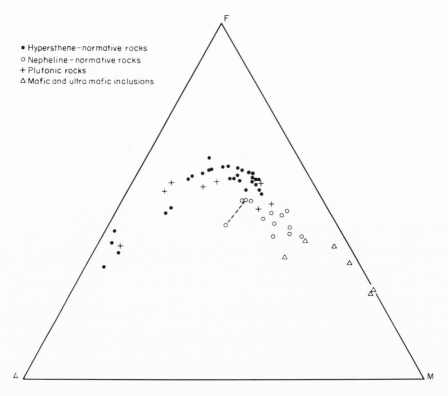

Figure 27. AMF diagram showing relations of analyzed rocks. $A = Na_2O + K_2O$; $M = MgO$; $F = FeO + 0.9\ Fe_2O_3$.

were erupted at a late stage, following a long period during which hypersthene-normative lavas were erupted. The widespread, youthful flows around James and Sullivan bays are very basic, nepheline-normative lavas. All of the picrite basalts that we saw in the Galápagos Islands were discharged at this late stage; all are rich in normative nepheline (for example, No. 76) and belong to the alkali-olivine basalt rather than the tholeiitic suite. Ankaramites appear to be totally absent from the archipelago.

The alkali-olivine basalts, taken as a group, exhibit a considerable range of composition, as may be seen in Figure 27. This range can be explained only in part by variations in the abundance of olivine. Compared with the tholeiites, the alkali-olivine basalts are further distinguished by their greater content of Ni and Cr.

Magnesia-Poor Alkali Basalts

These lavas are found in association with olivine tholeiites along the northeastern side of the archipelago, between Culpepper and eastern Chatham islands. Most of them are extremely rich in large phenocrysts of plagioclase but contain few of olivine. Aphyric types, although less conspicuous, commonly accompany strongly porphyritic types, even within individual flows (Pl. 20). The ground-mass of the porphyritic parts of the flows is chemically identical to that of the adjacent nonporphyritic parts; both are moderately alkaline, with approximately 2.5 percent normative nepheline and only 6 or 7 percent magnesia. We regard these rocks as differentiates of the olivine tholeiites with which they are closely associated. Analyses are presented in Table 2b (Nos. 4, 5, 5g, 8), and the average of Nos. 4, 5g, and 8 is given in Table 4. Trace elements of sample No. 4 are shown in Table 7. Petrographic descriptions are included in our accounts of Abingdon, Bindloe, Wenman, Culpepper, and Tower islands.

RELATIONS AMONG THE PRINCIPAL BASALTS

The principal types of Galápagos basalts—tholeiites, olivine tholeiites, and alkali-olivine basalts—are distinguishable by differences in MgO, TiO_2 and, to a lesser degree, by SiO_2 P_2O_5, and total alkalis (Table 4). Both tholeiites are distinctly lower than alkali-olivine basalts in both Na_2O and K_2O. All three averages are very close in total iron. Alkali-olivine basalts are much richer in Ni and Cr than are tholeiites.

Magnesium-poor alkali basalts (No. 5, Table 4) have a major and trace-element composition similar to that of the eastern tholeiites with which they are associated, but they are richer in alkalis and titania.

High titania commonly characterizes alkali basalts in other parts of the world, but the magnesium-rich types of the central and southern Galápagos Islands are much lower in this component than are tholeiites of the same or nearby islands. This difference results largely from the high titania content of iron-rich tholeiites, there being less difference in TiO_2 between alkali-olivine basalts and more basic tholeiites. In the eastern islands the relations are the reverse: low-MgO alkali basalts are slightly richer in TiO_2 than are tholeiites of the same islands.

Variations of iron and titania in the Galápagos basalts are remarkably sympathetic (Fig. 26). In alkali-olivine basalts, these components show no apparent relation to magnesium content, but in tholeiites both iron and titania increase as the content of magnesium diminishes.

The basalts of the Galápagos Islands are marked by wide variation in MgO and small variation in total alkalis (Fig. 27). In these respects they differ from the basalts of Hawaii. Alkali-silica ratios do not provide the clear distinction between hypersthene- and nepheline-normative compositions that has been utilized in comparing basalts in Hawaii. Several unoxidized Galápagos basalts containing normative hypersthene and some that contain normative quartz fall in the field of alkali basalts as defined by the empirical alkali-silica ratios of Tilley (1950, p. 41-42) and of Macdonald and Katsura (1964, p. 86-87). This relation is most pronounced in rocks with high iron-magnesium ratios.

The nature of the empirical alkali-silica line (A-B, Fig. 28) that separates Hawaiian tholeiites from alkali basalts does not seem to have been recognized, despite its inferred petrologic significance. The line is obviously related to the almost-parallel curve for the alkali-silica ratio of plagioclase (An-Ab). Considering first the felsic components, a rock is undersaturated and contains normative nepheline for any composition in which SiO_2 is lower than that required for feldspars of the appropriate composition. A similar relation applies to the mafic constituents. Hypersthene appears in the norm when the composition includes more available SiO_2 than that necessary for olivine (approximately 40 percent). Hence for natural rocks with both mafic and felsic components, the boundary between nepheline- and hypersthene-normative compositions must be a line parallel to the feldspar curve with an origin on the silica abscissa at a value determined by the ratio of mafic to felsic components. Since lime enters normative pyroxene and plagioclase according to its relation to Al_2O_3, it too has an effect, although a

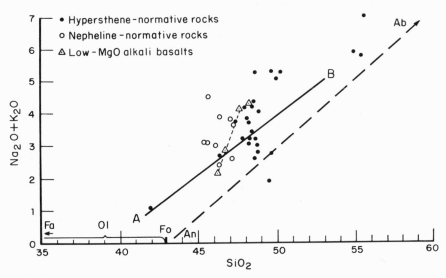

Figure 28. Alkali-silica diagram showing the relations of hypersthene-normative and nepheline-normative Galápagos rocks to the empirical boundary between the tholeiitic and alkaline basalts of Hawaii (line A-B). The broken line An-Ab shows the close parallelism of the alkali-silica values for plagioclase to the empirical boundary A-B.

subordinate one, on the origin but not on the slope of the boundary curve.

It follows, therefore, that the empirical boundary between alkali and tholeiitic basalts of Hawaii expresses a truism, because these rocks are defined according to their normative composition, which in turn is the basis of the dividing curve. In order to apply the empirical alkali-silica ratio to a wide range of rocks, adjustment of the origin of the curve must be made for different ratios of felsic and mafic components, but even if properly constructed for a particular suite of rocks, the line reveals nothing that is not already shown by individual norms, and the value of the diagram is limited to a graphic representation of a compilation of analyses and trends of differentiation.

Clearer distinction between the Galápagos basalts is obtained by comparison of their normative Di-Hy-Ol-(Q)-(Ne) contents, as plotted in Figure 29. Most analyzed tholeiites fall in a narrow field with a small amount of normative quartz, whereas alkali-olivine basalts form distinct groups with somewhat wider variations, especially in their content of normative nepheline. Low-magnesia alkali basalts fall outside the range of alkali-olivine basalts in an area of lower olivine but similar nepheline content. Olivine tholeiites containing normative hypersthene but neither nepheline nor quartz occupy a position between the quartz-normative tholeiites and alkali basalts.

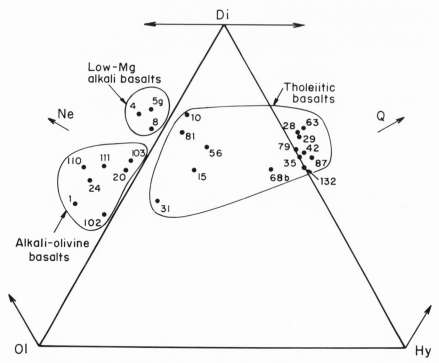

Figure 29. Relations of Galápagos basalts according to their ratios of normative diopside, olivine, hypersthene, nepheline, or quartz.

MAFIC AND ULTRAMAFIC INCLUSIONS

Coarse-grained inclusions already described from the basalts of Charles, Albemarle, Jervis, and James islands are of three kinds:

1. *Ultramafic rocks*, peridotites and dunites, found in alkali-olivine basalts on Charles Island.

2. *Gabbroic rocks*, chiefly eucrites, found among fragmental ejecta on Tagus and Beagle cones, Albemarle Island. Their host rocks are tholeiitic basalts, slightly enriched in iron.

3. *Differentiated plutonic rocks*, ranging from very basic gabbros to quartz syenites, found as inclusions in a pyroclastic cone on the northern shore of Jervis Island. These plutonic rocks, with the exception of the very basic types, correspond in chemical composition to the differentiated lavas of nearby vents on the same island. Gabbroic inclusions found at Buccaneer Cove on James Island resemble the mafic cumulates on Jervis Island.

Ultramafic inclusions have been found only with alkali-olivine basalts, gabbroic inclusions with tholeiitic basalts, and differentiated plutonic rocks with differentiated lavas.

Compositions of mafic and ultramafic inclusions and their clino-pyroxenes are presented in Tables 5, 6, 7 and 16. The plutonic rocks of Jervis and James islands are considered later, in the discussion of differentiated rocks (pages 148-152).

Peridotites from Charles Island are hercynite-bearing wehrlites, many of which contain minor amounts of calcic plagioclase, although none contains enstatite. These features distinguish the Charles Island inclusions from the more common enstatite-bearing peridotites of Hawaii and elsewhere that contain chromian spinel or chromite but no plagioclase. Coexistence of calcic plagioclase and magnesian

TABLE 5. CHEMICAL ANALYSES OF ULTRAMAFIC AND MAFIC INCLUSIONS

	64	74	98	110i
SiO_2	49.46	41.91	45.66	39.04
TiO_2	0.84	0.44	0.51	0.09
Al_2O_3	15.81	8.53	4.83	0.87
Fe_2O_3	1.52	3.21	1.58	1.19
FeO	5.73	11.25	6.68	13.69
MnO	0.17	0.16	0.18	0.28
MgO	9.38	28.43	25.57	44.36
CaO	14.50	4.96	13.21	0.04
Na_2O	1.80	0.99	0.62	tr.
K_2O	0.13	0.09	0.04	tr.
H_2O+	0.20	0.32	0.30	0.06
H_2O-	0.16	0.15	0.30	0.31
P_2O_5	0.03	..	..	..
Cr_2O_3	..	..	0.59	0.25
Total	99.73	100.44	100.07	100.18
Ap	0.05	..	..	..
Il	1.16	0.58	0.68	0.12
Or	0.80	0.50	0.25	..
Ab	16.15	8.40	1.00	..
An	34.60	17.77	9.80	0.20
C	..	..	Cm 0.93	0.77
Mt	1.59	3.17	1.57	1.13
Hm	..	..	..	Cm 0.37
Di	29.76	4.44	42.32	..
Hy	9.30	1.72	..	1.04
Ol	6.63	63.42	40.81	96.29
Q	..	..	..	..
Ne	..	..	2.58	..
Total	100.04	100.00	99.94	99.92
si	106.8	64.1	73.3	49.4
al	20.1	7.6	4.6	0.8
fm	42.3	82.8	71.6	99.1
c	33.6	8.1	22.7	0.1
alk	4.0	1.5	1.0	..
k	0.046	0.056	0.043	..
mg	0.720	0.781	0.858	0.772
qz	—9.2	—41.9	—30.7	—50.6

64. Gabbroic ejecta, Tagus Cone, Albemarle Island; host rock No. 63.
74. Olivine-augite eucrite block in pyroclastic cone on southern side of Buccaneer Cove.
98. Hercynite-chromian diopside peridotite inclusion, Charles Island; host rock No. 110.
110i. Dunite inclusion, Charles Island; host rock No. 110.

TABLE 6. Sr⁸⁷/Sr⁸⁶ and K/Rb Relations of a Charles Island Peridotite Inclusion and Its Host Lava

Specimen		K (ppm)	Rb (ppm)	Sr (ppm)	K/Rb	Rb/Sr	$(Sr^{87}/Sr^{86})_{meas.}$	(Sr^{88}/Sr^{86})	$(Sr^{87}/Sr^{86})_{corr.}$
Galápagos									
98	Peridotite	123	0.352	35.1	349	0.010	0.7041	8.385	0.7036
110	Host basalt	9460	58.9	435	161	0.135	0.7016	8.330	0.7035
Hawaii									
I-228i	Peridotite	....	....	....	....	....	0.7071	8.390	0.7064
I-228	Host basalt	....	....	....	....	....	0.7021	8.299	0.7054

98. Hercynite-bearing wehrlite, Charles Island. Major-element composition and analysis of clinopyroxene are given in Tables 5 and 16.

110. Alkali-olivine basalt, host of No. 98, Charles Island. Major-element composition and analysis of clinopyroxene are given in Tables 2c and 16.

I-228i. Chromian spinel-bearing wehrlite, 1801-1802 lava of Hualalai Volcano, Hawaii.

I-228. Alkali-olivine basalt, 1801-1802 lava of Hualalai Volcano, Hawaii.

olivine in the Charles Island wehrlites suggests that they represent a lower pressure assemblage than plagioclase-free lherzolites, but the significance of the difference in spinel compositions is not clear. There seems to be no petrographic evidence to suggest that plagioclase and hercynite were derived from reaction of magnesian spinel with diopside and enstatite. The analyzed clinopyroxene of the Charles Island wehrlite (No. 98, Table 16) is an aluminous chromian diopside. It resembles the chromian diopside of a peridotite (No. I-228, Table 16) collected from the 1800-1801 lava of Hualalai, Hawaii, but it is slightly higher in calcium, sodium, and aluminum. Both inclusions contain more than 50 percent SiO_2 and hence are more siliceous than the enclosing alkali-olivine basalts.

Dr. Alan M. Stueber determined the strontium isotopic composition of the analyzed peridotite from Charles Island (No. 98) and its host lava. The Sr⁸⁷/Sr⁸⁶ ratios of the two rocks are essentially identical (Table 6). For comparative purposes, a peridotite inclusion from Hualalai Volcano and its host lava were also analyzed. Although the isotopic ratios are slightly higher than those of the Galápagos rocks, the values for the inclusion and lava are essentially equal within the limits of the analytical technique (Table 6, Nos. I-228, I-228i). These relations seem to provide further evidence of the close genetic relation between peridotite inclusions and their host lavas. Differences in ratios of the Hawaiian and Galápagos rocks are of the same nature as those found among rocks from Ascension and Gough islands in the

Atlantic by Gast and his co-workers (1964), who interpreted the differences, together with variations in lead isotopic ratios, as reflections of small lateral variations in the composition of the upper mantle.

Both the Charles Island peridotite and its host lava have low K/Rb ratios (Table 6), similar to those found in other alkali-olivine basalts and inclusions. The lower ratio in the basalt is consistent with the predicted effect of fractionation resulting from the difference in ionic radii of the two elements and with the conclusion reached by Stueber and Murthy (1966) that peridotite inclusions represent refractory residua left by subtraction of basaltic liquid from a parent material with an initially low K/Rb ratio.

The uranium and thorium contents of the Charles Island peridotite (wehrlite) are unusually high. Values determined by duplicate neutron activation analyses by Wakita and his co-workers (1967) are:

Concentration ($\times 10^{-8}$ g/g), No. 98	
U	Th
19.8 ± 0.7	12.9 ± 0.3
19.5 ± 0.7	13.4 ± 0.3

The uranium value is about 8 times that of an enstatite-bearing peridotite from Hualalai Volcano in Hawaii, and the thorium value is about 17 times as high. The uranium and thorium contents of oceanic enstatite-bearing peridotites are variable but average about 5.48 and 2.14 $\times$ 10^{-8} g/g, respectively, but too few specimens have been analyzed to determine whether this difference can be correlated with the presence or absence of enstatite.

Eucrite inclusions in the tholeiitic basalts of Tagus and Beagle cones, Albemarle Island, are open-textured, meshlike assemblages of euhedral crystals with abundant interstitial glass. Some are crudely layered. The analyzed clinopyroxene (No. 64, Table 16) is a common augite. Compared with the augite (No. 228c, Table 16) of a gabbroic inclusion from the 1801-1802 lava of Hualalai Volcano, Hawaii, the Galápagos pyroxene has less aluminum and sodium but in other respects is quite similar.

Relations of Inclusions and Basalts

Interpretations of the origin and differentiation of Galápagos basalts are largely dependent upon the role assigned to ultramafic inclusions. One of the principal questions to be resolved is whether the different types of basalt could have been derived from a single parent or whether each is a distinct magma independently derived from a source in the mantle. Coarse-grained inclusions provide a means of testing the possibility of relating basalts through a mechanism of crystal fractionation.

One such relation can be examined by consideration of analyses of Galápagos tholeiites and their mafic inclusions. For example, subtraction of the mineral assemblage of a eucrite cumulate, such as No. 64, from a primitive tholeiite, such as No. 42, would yield a differentiate enriched in Ti, Fe, Na, K, Zr, and Cu and impoverished in Al, Si, Mg, Ca, P, Ni, and Cr. The ferrobasalt host (No. 63) of the analyzed eucrite is just such a rock. The relations of the major elements in these rocks is presented graphically in Figure 30a, from which it is readily apparent that subtraction of the eucrite-inclusion from the tholeiite (No. 42) produces a nearly rectilinear trend toward the composition of the host rock (No. 63). The graphical solution results in a predicted composition of the ferrobasalt that is within 1 percent of every major component in No. 63, and even the small discrepancies in Al_2O_3, CaO, and total FeO can be explained in terms of slight variations in the proportions of the mafic minerals and plagioclase in the coarse-grained layered assemblage that makes up the analyzed specimen.

This close relation between mafic cumulates and basic members of the tholeiite series is markedly different from the relations between alkali-olivine basalts and their ultramafic inclusions. To produce magnesia-rich compositions by fractionation of a primary tholeiite one must invoke resorption of magnesium-, nickel-, and chromium-rich crystals at hotter horizons. Even by this process, however, the relations of most other components cannot be accounted for by simple subtraction or addition. Figure 30a shows that a reversal of slope is required to join almost all major components; Al_2O_3 is the only major oxide lying on a rectilinear trend. A similar calculation, using an ultramafic inclusion such as No. 98 instead of a eucrite, shows equally well the impossibility of relating alkali-olivine basalts to tholeiites by separation of crystal cumulates (Fig. 30b). To overcome these difficulties by appeal to massive diffusion, the nature and volume of the components that must be moved become chemically and quantitatively inacceptable.

If one assumes a genetic relation between primary basaltic magmas and their characteristic mafic and ultramafic inclusions, limits can be placed on the chemical composition of a single parent rock from which all components of the two multiphase systems can be derived. The modal compositions of individual inclusions are too varied to permit calculations based on averages of small numbers of individual bulk analyses. In contrast to this diversity of modal composition, however, individual mineral components of the inclusions have a very limited range of composition. By using analyses

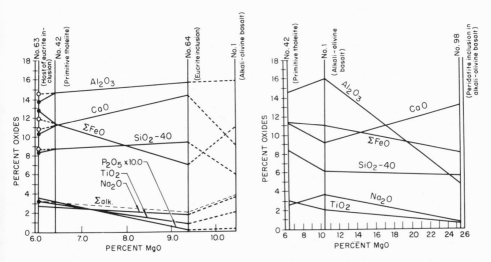

Figure 30. Relations of the major elements of Galápagos tholeiites and their mafic inclusions (a) and of the major elements of a typical Galápagos tholeiite, alkaline-olivine basalt and peridotie inclusion (b). (a.) Subtraction of the eucrite inclusion (No. 64) from a primitive tholeiite (No. 42) leads to a composition almost identical to that of the ferrobasalt that is the host lava of the eucrite inclusion (No. 63). (b.) The absence of any rectilinear relation between the three rocks shows the difficulty of relating one to the other by any simple process of subtraction or addition.

of these minerals, one can determine the possible effects of their subtraction from a parental composition. One method is illustrated in Figure 31. Successive pairs of oxides are plotted for the two basalt types, together with each of the constituent minerals found in their inclusions. When the three points representing the alkali-olivine-basalt liquid and the two minerals of its inclusions, diopside and olivine, are plotted, a triangular area (Di-D-A.O.) is described within which the composition of the parent of the two minerals and liquid must lie. Similarly, the quadrilateral area between the points plotted for the tholeiitic liquid (Th) and the augite (Au), olivine (Ol), and bytownite (By) of its eucrite inclusion delimits the range of the two oxide components for the parent of the tholeiitic system. The area shared by these two systems (Di-X-Y) represents the composition of a parent rock from which both basaltic liquids could be derived by subtraction of the minerals of their inclusions.

More than fifty possible pairs of the major elements can be tested in this fashion, but only a few vary enough to be useful in delimiting the possible range of parental compositions. In all cases, there is a common field of overlap between the alkali-olivine basalt and tholeiitic systems. By using the minimum range of values satisfying both systems for each oxide, the range of each component is narrowly limited. When one calculates the proportions of the mineral components that must be subtracted from the resulting composition, the proportions of crystals and liquid differ for the various oxide ratios by an amount that exceeds the possible values allowed by the graphical solution. It must be concluded, therefore, that the genetic relation between the lavas and inclusions may not be that of a simple closed system in which liquids were extracted from a homogeneous parent leaving residua of refractory minerals. An alternative explanation is that both basalts were in equilibrium with rocks that differed not only in physical conditions but in chemical composition as well.

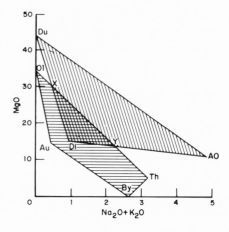

Figure 31. Method of determining graphically the limiting range of a single parent from which both tholeiites and alkali-olivine basalts could be derived by subtraction of the mineral components of related inclusions. (*See* text for detailed explanation.)

FRACTIONATION OF CALCIC PLAGIOCLASE

Low-magnesia alkali basalts, such as those found in the northeastern islands of the archipelago, do not present such serious obstacles to differentiation from their associated tholeiites because they are enriched in alkalis without being significantly richer in magnesia. Moreover, the abundance of calcic plagioclase phenocrysts in many of these lavas indicates that they may not be primary magmas but products of crystal fractionation.

Similar lavas are to be found in many volcanic associations elsewhere in the world. Kennedy (1931), for example, has described feldspar-phyric basalts from Scotland. There, as in the Galápagos Islands, they occur in composite flows with aphyric alkalic rocks having the same composition as the groundmass of the porphyritic rocks. Macdonald and Katsura (1964) found that feldspar-phyric basalts of alkalic character were erupted during the transitional stages of growing Hawaiian volcanoes, where they are associated with both tholeiitic and alkali-olivine basalts, appearing to mark a stage in the evolution toward succeeding alkalic differentiates.

Tilley and others (1965) investigated the melting relations of one of the Hawaiian feldspar-phyric basalts. Their specimen contains fewer phenocrysts than the analyzed Galápagos lavas and, despite its alleged position in the alkali series, it contains more than 5 percent normative hypersthene. They interpret the rock as a differentiate from alkali-olivine basalt by flotation of plagioclase and sinking of olivine. From calculations of density relations, they deduced that plagioclase phenocrysts would float in basaltic liquid so as to accumulate in the upper part of a magma reservoir, while olivine would settle and so be lost to the liquid erupted as lava. Their calculations appear to have been based on density relations at room temperature and thus neglect the important factor of thermal expansion, which is significantly greater for basaltic liquids than for crystalline plagioclase. It is difficult to decide on the basis of available data whether or not this factor would be great enough to reverse the density relations. Judging from large layered intrusions, it would appear that basic plagioclase sinks in liquids of basaltic composition, but studies of the Hawaiian lava lakes suggest that at least under some conditions the crystals show little tendency to sink or float (Peck and others, 1966).

The normative composition of the phenocrysts, calculated from the difference between the analysis of the total rock (No. 5) and its groundmass (No. 5g), consists of 93.5 percent feldspar ($An_{80.0}Ab_{18.9}Or_{1.1}$), 3.8 percent olivine, 0.8 percent magnetite, 0.3 percent diopside, and 0.9 percent nepheline. Subtraction of this composition from a tholeiitic magma would give a product much lower in Al_2O_3 and CaO and higher in SiO_2 and MgO than the aphyric alkali basalt in the composite flows. Subtraction of olivine as well would reduce the discrepancy in MgO but would also cause an increased iron-magnesium ratio and a greater difference in SiO_2. In short, we can suggest no simple mechanism of crystal fractionation by which low-magnesia alkali basalts can be produced from either of the other principal basalts. If they are genetically related to the tholeiites with which they are associated, some other mechanism of differentiation or fractionation of another combination of minerals must have operated.

TRACE ELEMENTS

Dr. S. R. Taylor determined the concentration of twelve trace elements in a selected group of Galápagos rocks. The analyses were carried out in Canberra on a direct-reading spectograph calibrated against interlaboratory standards, such as W-1 and G-1. Results are tabulated in Table 7. Variations of individual components as a function of major-element compositions of differentiated lavas and plutonic rocks are shown in Figures 32 and 33. Concentrations (in parts per million) are plotted against an index of differentiation, S.I., which is essentially the "Solidification Index" of Kuno (1962) but with the minor difference that all iron is calculated as FeO in order to reduce the effects of varying degrees of oxidation. Hence,

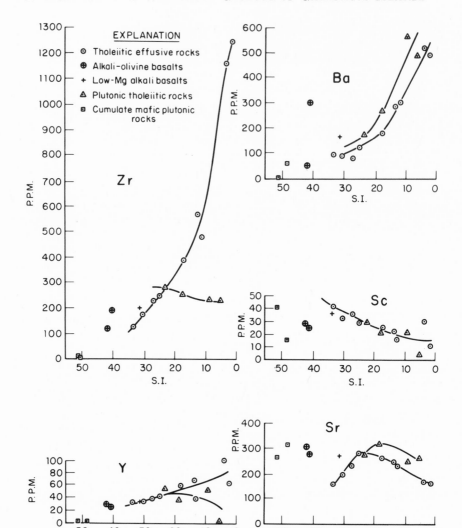

Figure 32. Variations of Zr, Ba, Sc, Y, and Sr as a function of the "Solidification Index" of Galápagos rocks.

$$\text{S.I.} = \frac{\text{MgO} \times 100}{\text{MgO} + \text{FeO} + 0.9\,(\text{Fe}_2\text{O}_3) + \text{Na}_2\text{O} + \text{K}_2\text{O}}$$

Trace elements of ultramafic inclusions are not shown in Figures 32 and 33, because for most elements they differ too widely from the concentrations in other rocks to be plotted accurately on the same scale.

The trace elements of the Galápagos rocks, when compared with those of rocks from other regions, show broadly similar concen-

	Tholeiitic Basalts			:ks		Mafic and Ultramafic Inclusions		
	35	42	63	67-D	67-N	64	98	110i
SiO_2	48.68	48.79	48.45	55.39	62.12	49.46	45.66	39.04
TiO_2	1.96	3.02	3.39	1.52	0.80	0.84	0.51	0.09
Cr_2O_3	..	..	..	..	..	..	0.59	0.25
Al_2O_3	15.22	14.72	13.75	15.02	18.16	15.81	4.83	0.87
Fe_2O_3	3.71	2.80	4.72	7.35	2.01	1.52	1.58	1.19
FeO	7.43	8.91	8.60	4.79	3.50	5.73	6.68	13.69
MnO	0.20	0.16	0.20	0.17	0.08	0.17	0.18	0.28
MgO	7.01	6.43	6.05	1.92	0.82	9.38	25.57	44.36
CaO	11.72	11.24	10.71	5.05	3.17	14.50	13.21	0.04
Na_2O	2.55	2.62	2.79	7.78	6.45	1.80	0.62	tr.
K_2O	0.34	0.42	0.50	1.35	1.49	0.13	0.04	tr.
H_2O+	0.66	0.39	0.36	0.64	0.68	0.20	0.30	0.06
H_2O-	0.09	0.01	0.02	0.13	0.06	0.16	0.30	0.31
P_2O_5	0.24	0.31	0.36	0.64	0.17	0.03	..	..
Total	99.81	99.82	99.90	99.75	99.51	99.73	100.07	100.18
SI*	33.9	30.7	27.3	9.4	5.8	51.0	74.5	75.2

Ba	95	92	76	560	480	..	..	..
Sr	160	220	235	245	270	265	..	..
La	..	..	..	..	..	..	..	..
Y	33	34	40	54	..	..	..	..
Zr	125	175	230	235	230	..	..	..
Mn	1350	1200	1440	2600	800	870	870	1700
Cu	45	100	150	27	9	45	48	9
Co	47	46	55	10	5	150	>1000	>1000
Ni	90	68	70	<3	<3	46	100	190
Sc	42	33	37	23	5	42	48	3
V	270	280	310	23	8	120	160	14
Cr	270	210	47	<5	6	220	>500	>500

Cs	≤1.5	..	..	1.2	1.1	..	..	..
La	9.0	12.5	15.8	42.5	21.0	5.7	2.4	0.6
Sm	3.6	4.8	5.7	11.5	3.2	1.6	0.4	..
Eu	1.5	..	..	4.6	3.9	..	..	..
Hf	1.4	..	4.1	4.8	7.6	≤3.0	≤3.0	..
Co	41.0	48.0	55.0	11.3	6.6	52.0	86.0	163.0
Sc	43.2	40.8	45.3	21.7	6.4	44.4	57.0	2.5
Cr	305	479	142	..	..	..	4060	3410

Ba	95	90	65	560	505	..	15	..
Sr	190	325	323	277	290	..	30	..
Zr	127	182	255	221	346	..	13	..
Nb	17	22	22	29	14	..	5	..

*SI (Solidification Index of

†Same analyses are presente(

‡Determinations by S. R. Tastralia. Detection limits in parts per
million: Ba, 20; Sr, 10; La

§Determinations by G. G. G(Cs, 20 to 50; La, 3 to 10; Sm, 3 to 8;
Eu, 2 to 3; Hf, 2 to 10; C(

**Determinations by E. Kabl

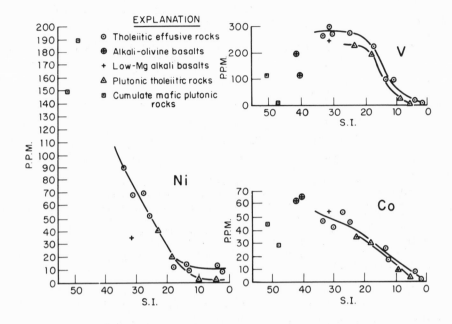

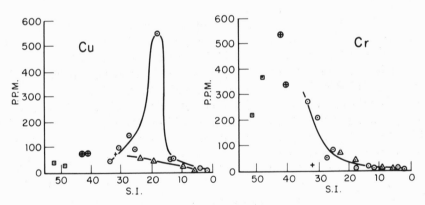

Figure 33. Variations of Ni, V, Co, Cu, and Cr as a function of the "Solidification Index" of Galápagos rocks.

trations and trends of differentiation. Notable exceptions are Ni, Cr, and Sr (Table 8). Both Ni and Cr show a greater contrast between Galápagos tholeiites and alkali-olivine basalts than they do in corresponding basalts from Hawaii, as might have been anticipated from their relative contents of Mg. Many Hawaiian tholeiites are richer in Mg than are common Hawaiian alkali basalts; they are also richer in Ni and Cr. The reverse is true of the Galápagos lavas.

Table 8. Comparison of Hawaiian and Galápagos Tholeiites
and Alkali-Olivine Basalts*

| | Tholeiites | | Alkali-Olivine Basalts | |
	Galápagos†	Hawaii‡	Galápagos§	Hawaii**
Mg	38.5 • 10³	44 • 10³	62.7 • 10³	32 • 10³
Ni	68	200	>400	65
Cr	210	300	345	250
Ca	80.3 • 10³	71 • 10³	65.7 • 10³	81 • 10³
Sr	220	350	280	1000

*Concentrations of Mg, Ni, Cr, Ca, and Sr in parts per million.
†Southwestern slope of Narborough volcano, elevation 1800 feet (No. 42).
‡Hypersthene-bearing basalt, 1887 lava, Mauna Loa (Nockolds and Allen, 1954, Table 29, No. 6).
§Alkali-olivine basalt between laboratory and dock, Darwin Station, Academy Bay, Indefatigable Island (No. 1).
**Olivine basalt, northeastern side of Mauna Kea (Nockolds and Allen, 1954, Table 1, No. 2).

The trace-element composition of the analyzed low-magnesia alkali basalt from Abingdon Island (No. 4) shows a close affinity to tholeiites, such as No. 35 (Table 8), with which it is closely associated; however, Ni and Cr seem to be depleted disproportionately in relation to Mg.

Strontium is much less abundant in Galápagos basalts than it is in comparable rocks of most other regions; it is more abundant in alkali-olivine basalts than in tholeiites, but its concentration shows no consistent relation to Ca. It is moderately enriched in the middle stages of differentiation of tholeiitic rocks (Fig. 32).

A striking feature of the trace-element variation is the very strong enrichment of Zr in highly differentiated plutonic rocks (Fig. 32). Copper is greatly enriched in a ferrobasaltic lava, but the plutonic series shows a steady decline in Cu concentration with progressive differentiation (Fig. 33).

PALAGONITIC BASALTS

Early theories regarding the origin of palagonite tuffs were based in large part on studies of material from the Galápagos Islands, where Darwin first described such rocks. Spectacular exposures in wave-cut littoral cones throughout the archipelago provide excellent opportunities for collecting samples and for observing field relations, some of which have already been discussed in our accounts of individual islands.

Darwin was intrigued by the distinctive orange-brown fragmental debris in the littoral cones of Chatham, James, and Albemarle islands and was probably the first to recognize a relation between this palagonitic material and eruptions through water. Shortly thereafter, in 1851, Bunsen studied the chemical character of palagonites and published an analysis of a specimen from the Galápagos Islands that clearly demonstrated its hydrated character. Washington and Keyes (1927), after examining specimens from Eden Islet, collected by William Beebe, concluded that the tuffs are products of hydration and alteration of basaltic glass.

Palagonitization of basalt is now recognized to be intimately related to eruptions through an aqueous environment, but the process by which the alteration of the hydrated glass comes about is still being discussed. Some workers maintain that the alteration is independent of weathering and is a direct result of interaction between basaltic liquid and water at the time of eruption; others assign an important role to subsequent alteration, by either hydrothermal processes or weathering.

Something more than rapid quenching of basaltic liquids seems to be required. When this is done in the laboratory, the basalt shatters into fresh, homogeneous, glassy fragments. Many lavas that have poured into the sea are similarly unaltered. We have described discordant relations between zones of palagonitization and layering of tuffs on Daphne Island (p. 8). These crosscutting relations clearly show that alteration of the glass took place after the tuffs were deposited. It is not evident, however, whether the alteration was caused by surface weathering or by steam and hot solutions permeating particular parts of the porous tuffs. Differences in physical or chemical conditions along the gradient of the rising and outwardly migrating solutions might play a part.

In an attempt to throw light on these possibilities, we subjected samples of natural basaltic glasses to a range of hydrothermal conditions that must encompass the possible extremes of temperature and pressure of natural solutions streaming through a pyroclastic cone. Two samples were used. The first was a tachylitic glass from the 1960 Kapoho lava of Hawaii, collected from the beach where the lava entered the sea; the second was a remarkably clear sideromelane from the glassy selvage of a pillow dredged from the ocean floor on the East Pacific Rise (No. 1, Table 20). The samples were heated in both distilled water and artificial sea water for periods up to six weeks and at temperatures and water pressures up to 500°C and 300 bars.

Hydration of the glass was accomplished with little difficulty, but in no case did we form a product even remotely similar to palagonite. Rims of low refractive index were produced on clear glass in a few hours, but a maximum effect was quickly reached beyond which no further lowering of refractive indices was observed.

This is shown in Figure 34 in which the lowest observed index is recorded for samples held at 350°C and 125 bars water pressure for periods ranging from about an hour to 8 days. The temperature and pressure selected for this series of runs were those at which a maximum lowering of refractive index was observed in 4-day runs over a range of temperature and pressure. At higher water pressure, the glass was found to devitrify rapidly; at higher temperatures magnetite crystallized and the glass became opaque. At lower temperatures and pressures, the lowering of refractive indices was less marked. No difference was noted between runs with distilled water and artificial sea water. Abundant analcite and phillipsite were present in samples showing the greatest alteration.

From these results we conclude that hydrothermal alteration of hydrated basaltic glass does not produce typical palagonite and hence that slower, low-temperature weathering processes must play a dominant role. The role of an aqueous environment is essential in producing the original basaltic glass, but the main effect of the initial hydration of the glass must be to accelerate subsequent alteration to palagonite by low-temperature processes.

The chemical effects of palagonitization of a Galápagos basalt have been studied by De Paepe (1966). His analyses of sideromelane and palagonite from Daphne Island are given in Table 9 together with analyses of similar material presented by Washington and Keyes (1927) and Bunsen (1851).

De Paepe concluded that palagonitization of the Daphne Island tuffs resulted in a decrease of SiO_2, MgO, and CaO and an increase

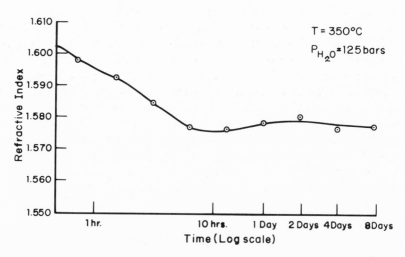

Figure 34. Lowest observed refractive indices of hydrated rims on fragments of basaltic glass held at 350°C and 125 bars for periods up to 8 days. The original refractive index of the unaltered glass was approximately 1.605. Note logarithmic time scale.

TABLE 9. ANALYSES OF SIDEROMELANE AND PALAGONITE

	1	2	3	4	5	6
SiO_2	48.00	43.45	44.75	45.35	38.13	38.07
TiO_2	0.70	0.80	0.70	0.85	2.50	n.d.
Al_2O_3	16.30	18.45	16.25	17.10	14.64	13.03
Fe_2O_3	2.80	5.90	5.95	4.65	7.93	9.99
FeO	7.60	5.60	6.30	6.60	0.87	..
MnO	0.10	0.15	0.15	0.15	0.15	n.d.
MgO	10.70	7.50	9.00	9.05	3.84	6.58
CaO	10.10	9.00	8.75	9.85	8.97	7.54
Na_2O	2.55	2.35	2.40	2.50	2.67	0.70
K_2O	0.15	0.45	0.35	0.40	0.15	0.94
H_2O+	0.65	2.90	3.70	2.40	12.34 ⎫	23.14
H_2O-	0.65	2.80	1.50	0.95	8.41 ⎭	
P_2O_5	0.45	1.00	0.80	0.70	0.01	n.d.
Total	100.75	100.35	100.60	100.55	100.61	99.99

1. Sideromelane from Daphne Island (De Paepe, 1966)
2. Palagonite tuff, Daphne Island (De Paepe, 1966)
3. Palagonite tuff, Daphne Island (De Paepe, 1966)
4. Palagonite tuff, Daphne Island (De Paepe, 1966)
5. Palagonitic tuff, Eden Islet (Washington and Keyes, 1927)
6. Palagonite (Bunsen, 1851)

of H_2O, P_2O_5, and the ratio of ferric to ferrous iron. He was particularly impressed by the strong enrichment of P_2O_5, which does not seem to have been reported by other workers. He does not indicate to what extent the phosphorus may have been derived from guano. Daphne Island is a nesting place for thousands of frigate birds and boobies.

DIFFERENTIATED ROCKS

Perhaps the most interesting petrologic feature of the Galápagos volcanoes is the wide and continuous range of compositions among the differentiated effusive and plutonic rocks produced during late stages of activity on some of the islands in the central part of the archipelago. We have already described the petrographic character of these rocks in our discussion of Jervis, James, and Duncan islands.

Volume and Distribution

The total volume of differentiated rocks is very small, certainly less than 1 percent of the total mass of volcanic rocks that are now exposed. Although exact measurements are impossible, the relative

volumes of successive members of the effusive series become progressively smaller with increasing content of silica. Intermediate
rocks (50 to 59 percent SiO_2), although minor in amount compared
with common basalts, constitute substantial parts of Duncan Island
and most of Jervis. The most siliceous compositions (60 percent
SiO_2 or higher) are represented by only one or two lavas and a
single bed of pumice.

Jervis Island is composed almost wholly of differentiated lavas,
and nearly all of the plutonic rocks that we found came from among
the ejecta of a littoral cone on its northern shore. Duncan Island
probably has a comparable volume of differentiated lavas, but we
were not able to examine the whole island in detail, and we doubt
that our samples represent the full range of rock types that must
be present. Many late lavas of the early cycle of activity on James
Island were moderately differentiated. We could not locate the
trachyte collected there by Darwin; most of the specimens we
examined are much more basic. Coarse-grained rocks from the
pyroclastic cone bordering Buccaneer Cove, near the western end of
the island, are cumulates of very basic composition, mainly eucrites,
there being no intermediate or strongly siliceous plutonic rocks
similar to those on Jervis Island.

Principal Members

Effusive Rocks. Differentiated tholeiitic lavas fall into three
groups: (1) basalts greatly enriched in iron, (2) intermediate rocks
in which there has been marked enrichment in silica and alkalis,
and (3) highly differentiated rocks of low color index. Characteristic
features of each of these three types are summarized below; chemical
analyses of typical specimens are given in Table 10, and petrographic
descriptions appear on pages 127-129.

FERROBASALTS

Ferrobasalts, as the name implies, are lavas marked by
strong relative and absolute enrichment in iron. Total iron normally
exceeds 12 or 13 percent, and MgO is less than 6 percent. Silica,
which shows little enrichment with respect to primitive tholeiites,
ranges from about 48 to 50 percent. Phenocrysts are mainly plagioclase, and this is largely labradorite, but andesine may be present
in the groundmass and anorthoclase is a common late-crystallizing
mineral. Corroded olivine is not uncommon in the groundmass, but
the main mafic minerals are augite, magnetite, and ilmenite. Hypersthene is rare. An average of twelve analyses is shown in Table 11.

ICELANDITES

Icelandites are intermediate between ferrobasalts and trachytes. We have adopted the name by which Carmichael (1964) distinguished certain intermediate lavas in Iceland from the more aluminous andesites of continental orogenic regions, but we extend the name to rocks of somewhat lower silica content than did Carmichael in order that the divison between ferrobasalts and icelandites may correspond approximately to the point at which enrichment in alkalis and silica becomes dominant over enrichment in iron. Consequently, total iron in our icelandites is less than the 12 or 13 percent found in ferrobasalts and it decreases rapidly with increasing silica content. Nevertheless, the iron content of icelandites is nearly twice that of common orogenic andesites of comparable silica content, and alumina is markedly lower. These relations are shown by the comparative analyses in Table 12. Some Hebridean craignurites, leidleites, and inninmorites are chemically similar to the Galápagos icelandites, but we have not used these older names because their definitions depend more upon petrographic characteristics than upon distinctive chemical compositions.

Galápagos icelandites are characterized petrographically by abundant phenocrysts of plagioclase, usually labradorite zoned to andesine, and a few of pigeonite. Olivine seems to be absent both in the groundmass and as phenocrysts, although a few xenocrysts, probably derived from coarse-grained cumulate rocks, were observed in some specimens. The groundmass pyroxenes include both subcalcic augite and pigeonite, but hypersthene seems to be lacking.

SILICEOUS TRACHYTES

Siliceous trachytes are the most differentiated rocks of the Galápagos effusive series. Their silica content exceeds 60 percent and they contain less than 10 percent mafic minerals. Some specimens contain as much as 20 percent normative quartz. Similar silica-rich trachytes are common in the eastern Pacific (p. 175). Macdonald and Katsura (1962) referred to such rocks on Oahu as rhyodacites. Despite certain chemical similarities to calc-alkaline rhyodacites, the mineral composition, high sodium-to-potassium ratio, and genetic relations of the oceanic lavas require that they be distinguished from rhyodacites of orogenic continental type. Unfortunately, of the four Galápagos trachytes we studied, only one, that from Jervis Island, contains little glass and can be attributed to a known source. It contains phenocrysts of andesine and oligoclase, hastingsite, hedenbergite, and magnetite in a fine-grained groundmass of anorthoclase,

TABLE 10. CHEMICAL ANALYSES OF DIFFERENTIATED EFFUSIVE ROCKS (Continued)

	51	70	71	49	50	48	130
SiO$_2$	49.50	54.76	55.34	59.64	59.39	66.18	66.87
TiO$_2$	3.65	2.30	1.93	1.74	1.91	0.67	0.66
Al$_2$O$_3$	14.38	14.37	14.02	14.61	14.32	14.55	12.55
Fe$_2$O$_3$	5.51	4.43	3.31	4.28	2.60	5.55	1.84
FeO	8.23	7.18	8.73	3.88	5.41	1.59	2.53
MnO	0.23	0.24	0.17	0.14	0.17	0.11	0.09
MgO	3.80	2.70	2.71	2.10	2.13	0.33	0.60
CaO	8.00	6.67	6.54	4.68	5.12	1.07	1.10
Na$_2$O	4.38	4.57	4.54	4.24	4.44	5.91	5.32
K$_2$O	0.97	1.38	1.33	1.96	2.40	2.91	3.08
H$_2$O+	0.63	0.79	0.64	1.67	1.68	0.96	4.66
H$_2$O—	0.24	0.20	0.18	0.29	0.18	0.26	0.33
P$_2$O$_5$	0.75	0.69	0.65	0.47	0.51	0.11	0.05
Other	..	..	..	..	..	..	..
Total	99.97	100.28	100.09	99.70	100.26	100.20	99.68
Ap	1.63	1.46	1.65	1.01	1.09	0.24	0.11
Il	5.22	3.28	2.74	2.50	2.72	0.94	0.98
Or	5.90	8.40	10.00	11.05	14.55	17.45	19.25
Ab	40.40	42.00	41.55	39.40	40.90	53.80	50.40
An	17.15	14.92	13.20	15.65	12.28	4.52	1.38
C	..	..	..	..	..	..	..
Mt	5.89	4.74	3.51	4.64	2.77	2.34	2.03
Hm	..	..	..	..	..	2.35	..
Di	14.80	11.56	11.80	4.28	8.28	0.12	3.20
Hy	7.74	7.34	10.74	4.78	6.38	1.02	2.10
Ol	..	..	..	..	..	..	..
Q	1.29	6.24	4.82	15.64	10.99	17.12	20.55
Total	100.02	99.94	100.01	99.95	99.96	99.90	100.00
si	136.3	166.4	165.6	230.1	208.0	310.8	332.0
al	23.1	25.7	24.7	33.2	29.5	40.2	36.7
fm	40.2	36.2	38.1	31.3	30.9	18.8	22.2
c	23.4	21.7	20.9	19.4	19.2	5.4	5.8
alk	13.3	16.4	16.3	16.1	20.4	35.6	35.3
k	0.127	0.163	0.194	0.234	0.263	0.245	0.382
mg	0.387	0.340	0.319	0.389	0.361	0.125	0.202
qz	—16.9	+0.8	+0.4	+65.7	+25.4	+68.4	+90.8

51. Pigeonite icelandite on eastern side of crater, Jervis Island.

70. Icelandite, northern wall of caldera, Duncan Island.

71. Icelandite, eastern rim of caldera, Duncan Island.

49. Icelandite, blocky lava from dome near center of Jervis Island.

50. Porphyritic icelandite, lower southern slope of dome on Jervis Island.

48. Siliceous trachyte, small peak on southern side of south crater, Jervis Island.

130. Siliceous trachyte pumice, near coast at southeastern base of Alcedo Volcano, Albemarle Island.

TABLE 10. CHEMICAL ANALYSES OF DIFFERENTIATED EFFUSIVE ROCKS

	78	79	45	65	44	81	68
SiO_2	47.84	48.12	48.77	48.54	48.74	49.20	49.81
TiO_2	3.79	3.91	3.98	4.20	3.72	3.34	3.42
Al_2O_3	14.75	14.13	14.52	14.49	14.47	14.24	13.40
Fe_2O_3	5.01	4.21	4.38	4.67	3.74	3.92	7.22
FeO	7.85	8.96	9.20	9.16	10.07	9.76	7.80
MnO	0.19	0.19	0.22	0.16	0.17	0.17	0.24
MgO	5.21	5.51	4.66	4.58	4.01	3.86	3.61
CaO	9.22	9.59	8.72	8.22	8.52	7.97	7.92
Na_2O	3.63	3.00	3.29	3.58	4.30	4.45	4.20
K_2O	0.61	1.27	0.79	0.84	0.99	1.36	0.88
H_2O+	0.87	0.66	0.79	0.68	0.33	0.41	0.75
H_2O-	0.38	0.20	0.11	0.16	0.26	0.16	0.37
P_2O_5	0.43	0.46	0.55	0.47	0.70	0.92	0.59
Other	..	..	..	..	..	..	..
Total	99.78	100.19	99.98	99.75	100.02	99.76	100.21
Ap	0.94	0.98	1.20	1.00	1.50	1.98	1.28
Il	5.42	5.58	5.54	6.06	5.30	4.76	4.94
Or	3.70	7.90	4.85	5.15	6.00	8.25	5.45
Ab	33.60	27.65	30.55	33.30	39.45	40.90	39.15
An	22.68	21.85	23.55	21.70	17.63	15.20	15.68
C	..	..	..	..	..	..	..
Mt	5.37	4.51	4.72	5.04	3.99	4.19	7.85
Hm	..	..	..	..	..	..	..
Di	17.16	19.36	13.92	13.96	16.84	15.40	16.96
Hy	10.06	12.08	12.80	11.74	2.84	2.78	4.62
Ol	..	..	..	..	6.36	6.54	..
Q	1.11	0.01	2.76	2.05	..	..	4.18
Total	100.04	99.92	99.89	100.00	99.91	100.00	100.11
si	123.4	121.2	128.2	123.0	127.2	126.9	139.3
al	22.3	20.9	22.6	21.6	22.1	20.8	22.1
fm	42.3	43.9	43.2	45.9	41.6	43.8	41.3
c	25.4	25.9	24.5	22.3	23.8	22.0	23.7
alk	10.0	9.3	9.7	10.1	12.5	13.4	12.9
k	0.100	0.219	0.136	0.134	0.132	0.168	0.122
mg	0.477	0.473	0.423	0.379	0.377	0.341	0.365
qz	—16.6	—16.0	—10.6	—17.4	—22.8	—26.7	—12.3

78. Ferrobasalt exposed on beach north of Buccaneer Cove, James Island.
79. Ferrobasalt, accidental block ejected from pyroclastic cone south of Buccaneer Cove, James Island.
45. Ferrobasalt, thick coarsely crystalline lava on southern coast of Jervis Island.
65. Ferrobasalt, flow at eastern edge of lagoon on northern side of Jervis Island. Source near center of island.
44. Ferrobasalt with scattered plagioclase phenocrysts underlying 45 on Jervis Island.
81. Ferrobasalt, accidental block ejected from pyroclastic cone south of Buccaneer Cove, James Island.
68. Ferrobasalt, platy lava from southern side of caldera, Duncan Island.

clinopyroxene, magnetite, hematite, and glass. Abundant cristobalite lines the vesicles.

Plutonic Rocks. Differentiated coarse-grained plutonic rocks are of four types, only three of which have equivalent effusive rocks. Ferrogabbro is chemically similar to ferrobasalt, leucodiorite to icelandite, and quartz syenite to siliceous trachyte. No lavas are known however, to approximate the composition of various kinds of mafic cumulate rocks that we group together as eucrites. Chemical analyses and modal compositions are given in Tables 13 and 14; chemical variations are summarized in Figure 35; and petrographic descriptions are given on pages 43-44.

EUCRITES

Specimens vary markedly in modal composition, but all contain a large proportion of calcic plagioclase—usually bytownite, magnesium-rich olivine, and diopsidic augite. Magnetite, ilmenite, and apatite are conspicuously scarce. Some of the eucritic cumulate rocks of Jervis Island contain hornblende and biotite, but we found no rocks with significant amounts of enstatite or bronzite. The most distinctive textural feature of the eucrites is the poikilitic character of the clinopyroxene, which surrounds euhedral crystals of plagioclase and olivine. The bulk chemical composition of eucrites varies widely with the modal proportions of the minerals. Silica content ranges from 41.91 to 49.46 percent; MgO varies inversely with silica, from 28.43 to 9.38 percent.

FERROGABROS

These are more common among the ejecta on Jervis Island than are simple gabbros equivalent to the primitive tholeiitic basalts of Narborough and Albemarle islands. The analyzed specimens show

TABLE 11. AVERAGE COMPOSITION OF GALÁPAGOS FERROBASALT*

SiO_2	48.61
TiO_2	3.63
Al_2O_3	14.21
Fe_2O_3	4.66
FeO	8.97
MnO	0.20
MgO	4.87
CaO	9.12
Na_2O	3.56
K_2O	0.84
H_2O+	0.61
H_2O-	0.19
P_2O_5	0.53
Total	100.00
Ap	1.12
Il	5.18
Or	5.15
Ab	32.90
An	20.95
Mt	5.01
Di	17.76
Hy	11.14
Q	0.79
Total	100.00
si	120.7
al	20.8
fm	45.8
c	24.3
alk	9.1
k	0.135
mg	0.395
qz	—15.7

*Average of Nos. 28, 29, 44, 45, 51, 63, 65, 68, 68b, 78, 79, and 81.

TABLE 12. AVERAGE COMPOSITIONS OF INTERMEDIATE DIFFERENTIATED LAVAS
(ICELANDITES)

	1	2	3	4	5
SiO_2	55.05	59.52	59.27	59.77	61.89
TiO_2	2.12	1.87	1.12	0.77	0.68
Al_2O_3	14.20	14.46	13.68	17.58	17.07
Fe_2O_3	3.87	3.44	5.33	2.65	1.89
FeO	7.96	4.65	5.10	3.45	3.38
MnO	0.20	0.15	0.28	0.11	0.10
MgO	2.70	2.11	1.07	2.73	2.88
CaO	6.60	4.90	4.76	5.87	5.60
Na_2O	4.56	4.34	4.51	3.78	4.06
K_2O	1.36	2.18	2.00	2.03	1.81
H_2O+	0.71	1.67	1.42	0.71	0.45
H_2O-	0.19	0.23	1.39	0.13	0.16
P_2O_5	0.67	0.49	0.42	0.18	0.19
Total	100.19	100.04	100.35	99.76	100.16
Ap	1.44	1.05	0.90	0.38	0.40
Il	3.02	2.68	1.62	1.08	0.94
Or	8.25	13.30	12.35	12.25	10.75
Ab	41.90	40.15	42.35	33.50	36.55
An	14.73	13.98	11.78	25.90	23.10
Mt	4.12	3.72	5.82	2.81	1.98
Di	11.44	6.32	7.96	2.00	2.88
Hy	9.12	5.42	2.34	9.38	9.74
Q	5.98	13.38	14.88	12.70	13.66
Total	100.00	100.00	100.00	100.00	100.00
si	160.3	204.8	206.0	193.0	208.0
al	24.3	29.3	27.8	34.5	33.7
fm	39.7	33.5	35.0	29.3	29.0
c	20.5	18.0	17.7	20.4	20.2
alk	15.5	19.2	19.5	15.8	17.1
k	0.165	0.249	0.227	0.267	0.227
mg	0.296	0.325	0.159	0.453	0.500
qz	−1.7	+28.0	+28.0	+29.8	+39.6

1. Basic icelandite, average of Nos. 70 and 71.
2. Icelandite, average of Nos. 49 and 50.
3. Icelandite, Thingmuli Volcano, eastern Iceland (Carmichael, 1964, Table 4, No. 16).
4. Average of seven andesites of the Guatemalan Highlands.
5. Average of twenty Cascade andesites.

somewhat less absolute enrichment in iron than do the analyzed ferrobasalts, but they are relatively depleted in magnesium, so the iron-magnesium ratios are close to those of ferrobasalts. Plagioclase constitutes between one-half and two-thirds of the volume of the ferrogabbros, and it ranges in composition between labradorite and intermediate andesine. The clinopyroxene is a common augite. Olivine is present in all the specimens examined, but in some it is a magnesium-rich type (Fo_{75-80}) whereas in some of the andesine-bearing types it is distinctly richer in iron (Fo_{30-32}). Strongly pleochroic brown hornblende and ore minerals are abundant, and apatite is a common accessory. The texture is hypidiomorphic granular.

TABLE 13. Chemical Analyses of Plutonic Rocks

	74	67-O	67-C	67-F	67-L	67-J	67-I	67-D	67-E	67-N
SiO$_2$	41.91	46.96	47.04	47.25	48.02	48.30	50.15	55.39	55.60	62.12
TiO$_2$	0.44	1.96	0.71	1.63	2.30	2.77	2.95	1.52	1.59	0.80
Al$_2$O$_3$	8.53	15.25	22.24	15.26	15.19	15.88	16.43	15.02	15.14	18.16
Fe$_2$O$_3$	3.21	8.67	2.13	3.29	5.95	7.65	7.30	7.35	6.22	2.01
FeO	11.25	3.43	3.10	8.27	4.97	4.98	4.10	4.79	5.36	3.50
MnO	0.16	0.12	0.05	0.17	0.06	0.14	0.16	0.17	0.29	0.08
MgO	28.43	6.83	7.11	9.00	7.77	4.89	3.57	1.92	1.88	0.82
CaO	4.96	11.81	13.89	10.95	11.06	8.63	8.80	5.05	5.08	3.17
Na$_2$O	0.99	2.54	2.05	2.95	3.19	3.92	4.42	7.78	5.95	6.45
K$_2$O	0.09	0.21	0.55	0.23	0.66	0.51	0.95	1.35	1.36	1.49
H$_2$O+	0.32	1.64	0.69	0.44	0.75	1.31	0.56	0.64	0.38	0.68
H$_2$O—	0.15	0.58	0.19	0.12	0.13	0.19	0.04	0.13	0.14	0.06
P$_2$O$_5$	..	0.08	0.08	0.15	0.25	0.53	0.47	0.64	0.66	0.17
Total	100.44	100.08	99.83	99.71	100.30	99.70	99.92	99.75	99.65	99.51
Ap	..	0.16	0.16	0.32	0.54	1.14	1.02	1.34	1.42	0.33
Il	0.58	2.78	0.98	2.28	3.22	3.98	4.18	2.16	2.24	1.12
Or	0.50	1.30	3.25	1.35	4.05	3.10	5.75	8.15	8.20	8.83
Ab	8.40	23.20	17.22	26.60	29.65	36.40	40.90	53.05	54.20	58.05
An	17.77	30.18	49.77	27.82	24.80	25.00	22.55	11.30	10.75	14.78
C										0.53
Mt	3.17	9.03	2.21	3.45	6.25	6.00	3.42	7.83	6.57	2.10
Hm		0.12				1.50	2.90			
Di	4.44	23.68	14.76	20.68	23.08	12.64	15.08	8.16	8.36	..
Hy	1.72	7.72			1.60	7.96	2.86	1.82	3.46	5.30
Ol	63.42		11.00	17.50	6.81					
Q		1.83	Ne 0.65			2.28	1.34	5.19	4.80	8.96
Total	100.00	100.00	100.00	100.00	100.00	100.00	100.00	100.00	100.00	100.00
Si	64.1	102.2	104.2	101.0	107.4	120.5	130.5	166.7	168.0	236.5
Al	7.6	19.6	29.0	19.3	20.1	23.3	25.2	26.6	26.9	40.6
Fm	82.8	46.8	32.9	49.2	45.4	43.3	37.6	37.7	36.7	19.0
C	8.1	27.9	32.9	25.1	26.4	23.1	24.6	16.3	16.4	13.0
Alk	1.5	5.7	5.2	6.4	8.1	10.3	12.6	19.4	20.0	27.4
K	0.056	0.052	0.151	0.049	0.120	0.079	0.124	0.133	0.013	0.132
Mg	0.781	0.477	0.715	0.587	0.574	0.423	0.372	0.233	0.231	0.247
Qz	—41.9	—20.6	—16.6	—24.6	—25.0	—20.7	—19.9	—10.9	—12.0	+26.9

74. Olivine-augite eucrite, block in pyroclastic cone south of Buccaneer Cove, James Island.

67-O. Olivine-augite eucrite, block in pyroclastic cone on northern shore of Jervis Island.

67-C. Olivine-augite-hornblende-biotite eucrite, same locality as 67-O.

67-F. Olivine-augite-hornblende gabbro, same locality as 67-O.

67-L. Olivine-augite gabbro, same locality as 67-O.

67-J. Augite-hornblende ferrogabbro, same locality as 67-O.

67-I. Olivine-augite-hornblende ferrogabbro, same locality as 67-O.

67-D. Olivine-augite-hypersthene leucodiorite, same locality as 67-O.

67-E. Olivine-augite-hypersthene leucodiorite, same locality as 67-O.

67-N. Quartz syenite, same locality as 67-O.

LEUCODIORITES

Blocks of leucodiorite have a low color index, plagioclase making up about three-quarters of the volume. The plagioclase in two analyzed specimens is medium oligoclase with only slight normal zoning. Ferroaugite, ferrohypersthene, and fayalitic olivine are the chief mafic minerals. Biotite may be present in small amounts, but

Table 14. Modal Compositions of Analyzed Plutonic Rocks

	74	67-O	67-C	67-F	67-L	67-J	67-I	67-D	67-E	67-N
Plagioclase	36.1	60.9	67.9	50.1	62.3	51.9	59.0	75.2	68.2	90.5
Augite	6.8	31.9	16.3	20.2	20.0	18.5	19.6	11.9	9.1	4.7
Ca-poor pyroxene	..	..	..	..	..	..	..	1.0	5.1	0.4
Olivine	55.6	4.7	3.6	12.9	7.4	1.8	3.3	3.5	7.0	..
Hornblende	..	1.1	7.7	14.7	0.8	17.8	5.2	..	..	..
Biotite	..	..	4.2	..	..	..	tr.	..	0.1	..
Opaque oxides	1.5	1.3	0.3	2.0	9.5	8.6	10.8	6.4	7.3	3.0
Apatite	..	0.1	..	0.1	..	1.4	2.1	2.0	3.2	0.1
Quartz	..	..	..	..	..	..	..	..	..	1.1
Sphene	..	..	..	..	..	..	..	..	..	0.2
				Compositions of Minerals						
Plagioclase (An%)	82-78	81-60	81-74	55-43	57-43	48-35	78-23	21-15	20-16	18-14
Augite (Wo-En-Fs)	49-45-6	47-39-14	45-45-10	45-37-18	46-36-18	42-38-20	42-33-25	43-26-31	42-17-41	40-25-35
Orthopyroxene (Fs%)	..	..	..	22	22	31	..	73	73	72
Olivine (Fa%)	12	11	9	22	22	31	69	86	87	..

Note: Modal Ca-poor pyroxene includes uninverted pigeonite as well as hypersthene. Augite composition of 67-C, 67-J, 67-I, 67-D, and 67-N from chemical analysis. All others are optical determinations. Identification of specimens the same as in Table 14.

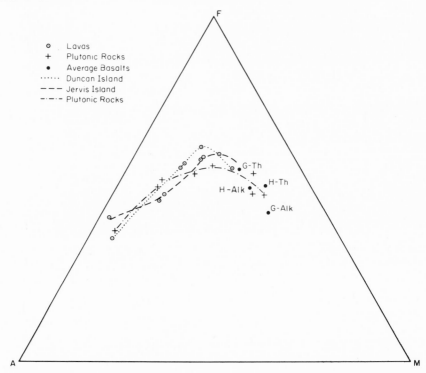

Figure 35. AMF diagram showing trends of differentiation in the lavas of Jervis and Duncan islands and in the plutonic rocks of Jervis Island. Points representing the compositions of average basalts of the Galápagos (G) and Hawaiian Islands (H) correspond to the averages of tholeiites (Th) and alkali-olivine basalts (Alk) in Table 4.

hornblende is either absent or very rare, its having been taken by ferrohypersthene. Opaque ores and abundant apatite make up the remainder.

Quartz Syenite

This, the most differentiated member of the plutonic series, is represented in our collection by only a single specimen. It consists of more than 90 percent potassic oligoclase, 4.7 percent ferro-augite, 3 percent opaque ore, about 1 percent quartz, and accessory ferrohypersthene, apatite, and sphene. The clinopyroxene is somewhat lower in iron than that in the leucodiorites, but the ferrohypersthene is richer in iron. Fayalite seems to be absent.

CHEMICAL AND MINERALOGIC EFFECTS
OF DIFFERENTIATION

The parent magma of the differentiated series was almost certainly tholeiitic basalt. No alkaline rocks have been recognized among the early lavas of the islands on which strongly differentiated rocks have been found. Alkali-olivine basalts that were erupted at a late stage on James Island represent a distinct new magmatic episode; no such rocks have been found on Jervis or Duncan islands.

Smooth and continuous chemical and mineralogic variations can be traced between basic members of the differentiated series and typical tholeiites of the large volcanoes of the western islands. In most variation diagrams, however, a significant gap and divergent trends separate the differentiated rocks and tholeiites from the compositional range of alkaline rocks.

Chemical Effects

These are characterized by strong iron-enrichment in early stages, followed in intermediate and late stages by rapid increase in silica and alkalis (Fig. 35). Even the most basic members of the sequence contain normative quartz; in the most siliceous rocks normative quartz reaches values as high as 20 percent. The rate of enrichment of alkalis relative to silica is intermediate between those of the alkaline and tholeiitic series of Hawaii. It closely resembles the trend for rocks of Easter Island (Fig. 41). Enrichment of potash with respect to soda is very slight in effusive rocks and virtually nil in plutonic rocks (Fig. 36). In even the most siliceous differentiates, soda is clearly the dominant alkali.

Titania increases rapidly with total iron in the more basic basalts, as shown in Figure 26. After reaching a sharp maximum near 48.5 percent silica, titania falls off rapidly with increasing silica content. (Fig. 37a). The titania content of plutonic rocks is uniformly lower than that of lavas, and mafic cumulate rocks (eucrites) are clearly distinguished by much lower titania than either lavas or plutonic rocks of similar silica content. The sympathetic variations of titania and iron closely resemble those observed in Alae lava lake, Hawaii (Peck and others, 1966). Phosphorus behaves in a similar manner (Fig. 37b). It increases sharply during early stages of differentiation and then drops off steadily. Plutonic rocks are lower in phosphorus than are lavas, and mafic cumulate rocks show extremely low concentrations of this component. The low phosphorus content of cumulate rocks is consistent with the mechanism of fractionation between adcumulates and liquids proposed by Wager (1960).

Trace-element variations have already been discussed (pages 137-140). These generally resemble variations found in differentiated sequences elsewhere. An interesting feature is the contrast between the strong enrichment in zirconium in differentiated lavas and the slight impoverishment in this element in corresponding members of the plutonic series.

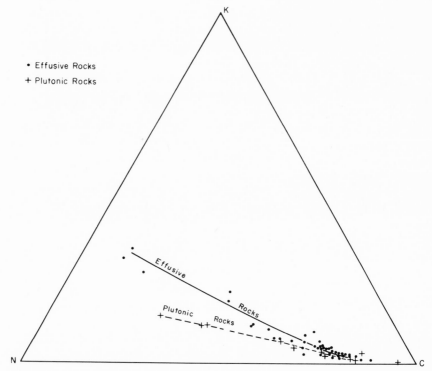

Figure 36. NCK diagram (Na_2O — CaO — K_2O) for analyzed Galápagos rocks.

Mineralogic Effects

These reflect chemical variations to a remarkable degree, particularly in the plutonic series. The effusive and plutonic rocks differ notably in the nature and regularity of their mineral sequences. Because of differing degrees of crystallinity displayed by the lavas and pyroclastic rocks, it is impossible to compare the proportions of the constituent minerals, particularly since nearly all of the specimens we studied contain important proportions of glass. Despite these textural differences, however, certain generalizations can be made.

Plutonic rocks, most of which are ejecta of a pyroclastic cone on Jervis Island, show various degrees of thermal alteration, and some have been partly remelted. Nevertheless, the original mineral and chemical compositions can be reliably estimated, because the specimens selected for analysis are only slightly altered or are quite fresh. The modal variations among the plutonic rocks are summarized in Table 14 on a glass-free basis.

Feldspars. Plagioclase was separated from each of the analyzed differentiated rocks and its composition was determined by the refractive index (x') of cleavage flakes. Compositions are shown in Figure 38 in comparison with normative compositions calculated from chemical analyses. A diagonal line joins points of equal modal and normative compositions.

Plagioclase becomes progressively more albitic in more siliceous rocks. It ranges from bytownite in eucrites and in the cores of large, zoned phenocrysts to albite in the groundmass of siliceous trachytes. The modal plagioclase of effusive rocks is consistently less albitic than that calculated from the norm, but in plutonic rocks the average modal plagioclase is almost exactly equivalent to the normative composition. Zoning is least pronounced in siliceous rocks. Anorthoclase is a late-crystallizing mineral in the interstices of almost all Galápagos lavas, but rarely does it constitute more than a very small fraction of the modal feldspar.

The only feldspar in the analyzed quartz syenite (No. 67-N) is a potassic oligoclase, the composition of which is of special interest since it is in an intermediate range not commonly found in coarsely crystalline rocks. A chemical analysis (Table 15) shows that this feldspar has the following molecular proportions: Or 9.0, Ab 75.9, An 15.1. The crystals are only weakly zoned and are twinned according to the albite and pericline laws. Properties are intermediate between those of common oligoclase and anorthoclase.

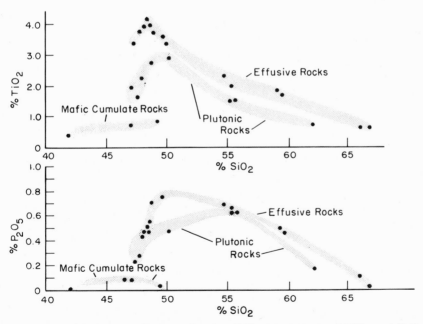

Figure 37. Variation diagram showing different contents of TiO_2 and P_2O_5 in effusive and plutonic rocks and in mafic cumulates as a function of silica content.

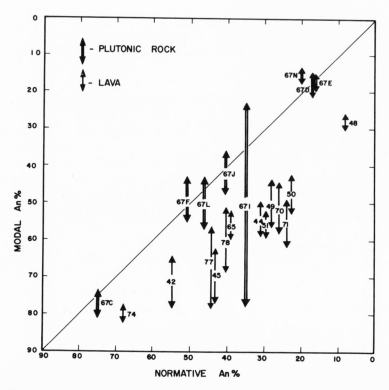

Figure 38. Comparison of modal and normative plagioclase compositions of differentiated lavas (solid bars) and plutonic rocks (open bars). The diagonal line indicates equal normative and modal compositions; the vertical range of the vertical bars shows the extent of zoning in modal crystals.

TABLE 15. CHEMICAL ANALYSIS AND PROPERTIES OF POTASSIC OLIGOCLASE
FROM QUARTZ SYENITE, 67-N

Weight Percent of Oxides		Atomic Proportions		Optical Properties	Molecular Proportions
SiO_2	63.61	Si	11.296		Or $=$ 9.0
TiO_2	0.00	Al	4.692	$N_x = 1.535$	Ab $=$ 75.9
Al_2O_3	22.47	Ti	0.000	$N_y = 1.540$	An $=$ 15.1
Fe_2O_3	0.33	Fe^{3+}	0.045	$N_z = 1.542$	
FeO	0.11	Fe^{2+}	0.016	$2V_x = 51—56°$	
MgO	0.02	Mg	0.005	$a':(010) = 12—16°$	
CaO	3.13	Ca	0.596	D $=$ 2.63	
Na_2O	8.75	Na	3.006		
K_2O	1.57	K	0.356		
	99.99				

Pyroxenes. The assemblages of mafic minerals change markedly with increasing differentiation. The most important of these minerals are calcium-rich clinopyroxenes, which are present in abundance in all rocks, both effusive and plutonic, and show a wide range of composition. Five of these clinopyroxenes were analyzed (Table 16). The compositions of other pyroxenes were determined from optical properties using the data of Hess (1949) and Brown and Vincent (1963). Optic angles were measured on a universal stage using crystals showing both optic axes in interference figures and were corrected for refractive index. Intermediate refractive indices were determined using well-formed single crystals mounted on a spindle stage (Wilcox, 1959). Compositions are plotted in Figure 39. Compositions determined from optical data are uniformly more calcic than those obtained by chemical analysis. The small difference probably results from submicroscopic exsolution of a calcium-poor pyroxene.

In effusive rocks, calcium-rich pyroxenes change from diopsidic augite in basic types to subcalcic augite in intermediate types. Only a slight amount of iron enrichment takes place throughout most of the course of differentiation, a trend similar to that pointed out by Kuno (1955). At an advanced stage of differentiation, however, the calcium-rich pyroxenes abruptly become richer in iron. Only two effusive rocks have been found that have a large hedenbergite component in their clinopyroxenes, and both of these rocks are siliceous trachytes (Nos. 48, 130).

In contrast to this trend in effusive rocks, the calcium-rich pyroxenes of plutonic rocks show a continuous series of progressively more iron-rich compositions. They follow a course similar to that noted among differentiated tholeiitic intrusive rocks in other regions.

Compositions of 116 published and unpublished analyses of calcium-rich clinopyroxenes from effusive and intrusive subalkaline rocks are plotted in Figures 40a and 40b, respectively. The trend toward subcalcic augite in all but the most differentiated effusive rocks and the scarcity of intermediate compositions are apparent. This trend contrasts markedly with that seen in differentiated intrusive rocks. The differences seem to us to reflect conditions of crystallization and to be a characteristic feature of subalkaline rocks.

Kuno (1955) suggested that substitution of Mg for Ca in subcalcic augite may result from minor substitution of Fe^{3+} for Si at high temperatures and consequent distortion of the pyroxene lattice. This explanation appears reasonable in the light of the conditions of crystallization of the Galápagos magmas inferred from other factors to be discussed below. Hedenbergite-rich pyroxenes and amphiboles present in the siliceous trachytes of the Galápagos Islands

TABLE 16. CHEMICAL COMPOSITIONS AND OPTICAL PROPERTIES OF CLINOPYROXENES

	98 Cr-Diopside	I-228 Augite	64 Augite	I-228c Augite	67-C Augite	67-J Augite	67-I Augite	67-D Fe-Augite	67-N Fe-Augite
SiO_2	50.25	50.54	50.95	48.78	51.87	51.30	51.12	50.72	51.29
TiO_2	0.58	0.61	1.14	0.94	1.01	0.73	0.53	0.48	0.50
Al_2O_3	6.03	4.41	2.84	5.11	3.98	1.86	1.27	0.59	1.36
Cr_2O_3	0.79	0.59	0.06	0.70	0.44	..	..	..	..
Fe_2O_3	1.62	2.29	2.64	3.16	1.61	2.48	2.34	1.83	1.92
FeO	3.12	4.80	6.58	4.64	4.52	9.79	12.36	17.00	17.24
MnO	0.16	0.16	0.23	0.15	0.16	0.16	0.51	0.76	0.74
MgO	15.25	16.41	15.36	14.68	14.56	13.02	11.36	8.74	7.72
CaO	20.61	19.05	19.97	20.50	20.32	19.88	20.03	19.71	17.68
Na_2O	0.79	0.65	0.42	0.63	0.57	0.63	0.30	0.47	1.12
K_2O	0.03	0.02	0.01	0.01	0.03	0.13	0.09	0.09	0.21
H_2O+	..	0.26	..	0.54	..	..	..	..	..
H_2O-	0.39	..	..	..	0.69	..	..	..	0.11
Total	99.62	99.79	100.20	99.84	99.76	99.98	99.91	100.39	99.89

Z

	98	I-228	64	I-228c	67-C	67-J	67-I	67-D	67-N
Si	1.848	1.861	1.885	1.818	1.915	1.929	1.948	1.965	1.990
Al IV	0.152	0.139	0.115	0.182	0.085	0.071	0.052	0.027	0.010
Fe^{3+}	..	..	..	..	..	..	..	0.007	..

WXY

	98	I-228	64	I-228c	67-C	67-J	67-I	67-D	67-N
Al VI	0.110	0.052	0.008	0.042	0.088	0.011	0.005	0.000	0.052
Ti	0.016	0.017	0.032	0.026	0.028	0.021	0.015	0.014	0.015
Cr	0.023	0.017	0.002	0.021	0.013	..	..	..	..
Fe^{3+}	0.045	0.063	0.073	0.089	0.045	0.070	0.067	0.046	0.056
Fe^{2+}	0.096	0.147	0.203	0.144	0.139	0.307	0.393	0.549	0.557
Mn	0.005	0.005	0.007	0.005	0.005	0.005	0.016	0.025	0.024
Mg	0.836	0.907	0.852	0.820	0.806	0.734	0.649	0.508	0.449
Ca	0.812	0.752	0.792	0.819	0.804	0.801	0.818	0.818	0.735
Na	0.056	0.046	0.030	0.046	0.041	0.046	0.022	0.035	0.084
K	0.001	0.001	0.000	0.000	0.001	0.006	0.005	0.005	0.010
WXY	2.001	2.007	1.999	2.012	1.970	2.001	1.990	2.000	1.982

	98	I-228	64	I-228c	67-C	67-J	67-I	67-D	67-N
Ca	45.1	40.2	41.1	43.6	44.7	42.0	42.1	42.5	40.4
Mg	46.8	48.3	44.2	43.7	44.8	38.5	33.4	26.4	24.6
Fe	8.1	11.5	14.7	12.7	10.5	19.5	24.5	31.1	35.0
$2V_z$	57.5°	54°	57°	58°	56°	49°	54.5°	52°	56°
Ny (±0.002)	1.695	1.697	1.695	1.690	1.687	1.698	1.699	1.713	1.713

98. Chromian diopside from hercynite-chromian-diopside peridotite, Charles Island. (Professor G. G. Goles has determined the following components by neutron activation analysis: Na $= 0.53 \pm 0.03$ percent, Sc $= 81.5 \pm 1.4$ ppm, Cr $= 0.733 \pm 0.015$ percent, Fe $= 5.1 \pm 0.5$ percent, Co $= 31.5 \pm 0.7$ ppm.)

I-228. Augite from peridotite inclusion, 1801 lava, Hualalai.

64. Augite from eucrite inclusion, Tagus Cone.

I-228C. Augite from gabbroic inclusion, 1801 lava, Hualalai.

67C. Augite from augite-biotite-hornblende-olivine gabbro, Jervis Island.

67J. Augite from ferrograbbro, accidental block, Jervis Island.

67I. Augite from ejected block, Jervis Island.

67D. Ferroaugite from leucodiorite, accidental block, Jervis Island.

67N. Ferroaugite from quartz-syenite block, Jervis Island.

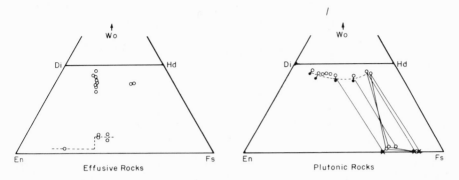

Figure 39. Compositions of pyroxenes in differentiated effusive and plutonic rocks. Open circles indicate compositions determined from optical properties, closed circles from chemical analyses.

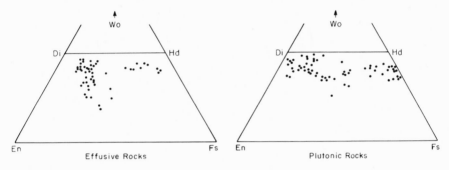

Figure 40. Compositions of analyzed calcium-rich clinopyroxenes from effusive (a) and intrusive (b) rocks.

are not considered normal members of the tholeiitic effusive series by Kuno (1959). However, Carmichael (1960) and others have described hedenbergitic pyroxenes in siliceous rocks from Iceland and Scotland; we believe, therefore, that Kuno's series should be amended to include these iron-rich compositions in extreme differentiates.

Carmichael's suggestion (1963) that magnesian pyroxenes in iron-rich volcanic rocks might result from early precipitation of magnetite is inadequate because it fails to explain the substitution of Mg for Ca. G. M. Brown (1966, personal commun.) suggests that rapid crystallization of the lavas causes precipitation of a pyroxene near the cotectic trough, corresponding to liquids poorer in calcium and magnesium than the corresponding equilibrium pyroxenes. This explanation seems reasonable, especially for variations in the compositions of pyroxenes in a limited range of basic and intermediate rocks. We note little consistent contrast, however, between phenocrysts and groundmass augites of the Galápagos lavas, and hypersthene and pigeonite coexisting with these augites show progressive iron enrichment. Neither of these features seems consistent with the quenching hypothesis. Although quenching effects could produce subcalcic augites, the compositional range in a series of progressively more iron-rich rocks should cover a broad field varying in iron-magnesium ratios as well as calcium content. This is not seen in the pyroxenes of Galápagos effusive rocks, which seem to form a very restricted trend with little if any variation in iron content. For these reasons we tend to prefer the explanation advanced by Kuno, even though it offers no explanation of the sudden shift to hedenbergite-rich pyroxenes in the most differentiated effusive rocks.

Despite the higher titania content of the tholeiitic ferrobasalts, there is no indication that their clinopyroxenes are significantly titaniferous. Almost all alkali-olivine basalts from the Galápagos Islands, although lower in titania than tholeiites, contain purplish-brown titaniferous clinopyroxenes. These relations suggest that incorporation of titania in clinopyroxenes is more dependent upon the degree of silica undersaturation and consequent high proportion of Al in z sites than on the availability of titania in the magma. This interpretation is supported by the fact that, even though titaniferous magnetite and ilmenite are abundant in all tholeiitic rocks and sphene is an important accessory in the most siliceous plutonic rocks, the titania content of the clinopyroxenes decreases with increasing silica content of the total rock.

Data concerning calcium-poor pyroxenes are scanty. Hypersthene is found in small amounts in only a few tholeiitic basalts and ferrobasalts, possibly because of the subcalcic nature of the augite. In icelandites, pigeonite takes the place of hypersthene but is never abundant. Plutonic rocks contain only minor amounts of pigeonite. Leucodiorites and quartz syenite (Nos. 67-D, 67-E, 67) contain ferrohypersthene that must be, at least in part, the product of inversion of ferropigeonite, which is also found in the same rocks. The preservation of pigeonite is probably the result of a very sluggish inversion at the low temperature at which these rocks must have crystallized (Brown, 1957).

Special mention should be made of a single crystal of augite, about 1.5 cm in diameter, that was collected by Dubois from the eastern part of Barrington Island, probably from a bed of palagonite tuff (p. 14). Large phenocrysts such as this are exceptionally rare in Galápagos lavas, although equally large phenocrysts of pigeonite are present in some of the icelandites of Jervis Island. A prominent feature of the Barrington Island augite is a strong zoning. Curiously enough, however, this zoning is best seen by the unaided eye as countless concentric shells. Under the microscope, the only difference that can be detected is a deeper brown tint in the marginal parts of the crystal. No significant differences in optical properties were noted from the core outward. The optic angle does not vary more than a degree from 55° in either the rim or the core; the refractive index is constant within the margin of error for our measurements, and no difference in dispersion can be seen. Traverses across the zones with an electron microprobe failed to reveal any difference in Ca, Mg, or Fe. Variations in Ti content seem to be ruled out by the constant 2V. We conclude, therefore, that the prominent brown rim probably results from variation in the oxidation state of titanium and iron during precipitation of the outer part of the crystal. Alternatively, oxidation may have resulted from reheating of the crystal during its eruption.

Olivines. Compositions of olivines have been determined from their optic angle.

Magnesian olivines are found in all basalts and ferrobasalts. In none of these rocks does olivine constitute a large proportion of the total volume, as it does in alkali-olivine basalts, and in none of the tholeiitic rocks does it occur in the groundmass. Reaction rims or other evidence of instability are not apparent, even in rocks with 1 or 2 percent normative quartz. We have not found fayalitic olivine in either of the two siliceous trachytes we collected, but the specimen reported to have been collected from James Island by Darwin contains rare microphenocrysts of fayalite. In contrast to the scarcity of olivine in differentiated effusive rocks, all plutonic rocks except the specimen of quartz syenite contain more than accessory amounts of this mineral. In basic plutonic rocks the composition varies from about Fo_{91} in eucrites to Fo_{75} in labradorite-bearing ferrogabbro, but an abrupt increase to hortonalite (Fo_{30-32}) occurs in the analyzed andesine-bearing ferrogabbro (67-I). Iron enrichment is further accentuated in leucodiorites, in which the olivine is ferrohortonalite (Fo_{13-15}).

Biotite. Biotite has been found in two Galápagos rocks, both plutonic ejecta from Jervis Island. In 67-C, an unusual rock probably of cumulate origin, biotite amounts to 4.2 percent of the volume and coexists with bytownite, diopsidic augite, hornblende, and olivine. The only other rock in which biotite has been observed is a single specimen of leucodiorite (67-E) in which it occurs as rare flakes too small to study more closely. The biotite of 67-C has a deep-brown color (x pale buff, y light tan, z dark brown) and moderately high refractive index ($N_y = 1.637 \pm 0.002$, $N_z = 1.641 \pm 0.002$). The axial angle varies from uniaxial to values as great as 10°. The variation probably reflects different degrees of oxidation resulting from reheating of the block by the enclosing basaltic lava.

Amphiboles. Hornblende is an abundant constituent of all but the most differentiated plutonic rocks. The chemical composition and optical properties of a typical amphibole, a titaniferous oxyhornblende from a ferrogabbro, 67-J, are given in Table 17. The high state of oxidation probably resulted from reheating by the basaltic lava in which the plutonic rocks occur as accidental ejecta.

The clinopyroxene coexisting with the oxyhornblende in 67-J has also been analyzed and is reported in Table 17. Compared with the amphibole, the pyroxene is notably more siliceous and has a slightly higher iron-magnesium ratio.

Ore Minerals. The principal ore mineral in both the alkaline and tholeiitic lavas is magnetite, ilmenite being much less common. Plutonic rocks collected from the pyroclastic cone on Jervis Island contain large proportions of hematite and maghemite which probably resulted from reheating of magnetite during eruption. The only coarse-grained rocks in which fresh magnetite was found without important amounts of hematite are the specimen of quartz syenite (67-N) from Jervis Island and a eucrite specimen from Tagus Cone on Albemarle Island (No. 64). The cell constant of magnetite from the quartz syenite (8.4407 ± 0.001) suggests that it is titaniferous, while that of the eucrite (8.412 ± 0.003) is less so.

The cell constant and presumably the titanium content of magnetites separated from lavas of a wide range of compositions show a crude relation with the TiO_2 content of the bulk rock (Table 18).

It has been shown (Aoki, 1966) that titanomagnetites of lavas may contain important amounts of spinel hercynite in solid solution, and these additional components must have an effect on the cell dimension. Moreover, magnetite from lavas with even small amounts of hematite were found to have anomalously small cell constants, and in the

TABLE 17. CHEMICAL COMPOSITION AND OPTICAL PROPERTIES OF OXYHORNBLENDE FROM FERROGABBRO, 67-J

SiO_2	43.40	$2V_x = 75°$
TiO_2	3.72	$N_x = 1.689$—pale brown
Al_2O_3	9.88	$N_y = 1.733$—dark brown
Fe_2O_3	11.07	$N_z = 1.759$—very dark
FeO	5.01	brown
MnO	0.32	$z \wedge c = 0°$
MgO	11.12	
CaO	11.96	
Na_2O	2.39	
K_2O	0.61	
H_2O+	0.84	
H_2O-	0.02	

TABLE 18. CELL CONSTANTS OF MAGNETITE FROM ROCKS OF VARIOUS TITANIA CONTENTS

Specimen Number	TiO_2 of Rock	a_o of Magnetite
1	2.07	8.409 ± 0.006
44	3.72	8.456 ± 0.005
78	3.79	8.446 ± 0.007
71	1.93	8.474 ± 0.002
48	0.67	8.412 ± 0.004
130	0.66	8.459 ± 0.002

Determinations by A. Pabst.

absence of analytical data it is impossible to draw further conclusions
on the significance of the ore-mineral compositions.

CONDITIONS AND MECHANICS OF DIFFERENTIATION

The coarse-grained plutonic rocks of Jervis Island provide a
valuable insight into the processes and conditions of differentiation
of the Galápagos igneous suite. Even though the genetic relations of
these rocks can only be inferred, they offer a means of relating the
tholeiitic differentiated series to an orderly mineralogical sequence.

We have already discussed the relation of eucrite inclusions to
basalts of the western and central islands (pages 133-136) and have
shown that although it is impossible to relate the alkali-olivine basalts
to tholeiites by crystal fractionation, separation of olivine, augite,
and basic plagioclase from the primitive tholeiitic magma could
have produced an iron-enriched differentiate such as the ferrobasalts
in which the eucrites are found.

Evidence of crystal fractionation is very convincing in these
mafic cumulate rocks; it is more difficult, however, to find evidence
to relate subsequent rocks to separation of later minerals, such as
amphiboles, magnetite, and intermediate pyroxenes and plagioclase.
Ferrogabbros have no clear cumulate textures and also differ from
more basic members of the series in being poor in olivine and rich in
amphibole and magnetite. In addition, their pyroxenes are more
iron-rich and their plagioclase more albitic. Subsequent differentiates, if they were derived from the ferrogabbro liquid by crystal
fractionation, must be related through separation of these minerals
rather than the more basic assemblage of eucrites.

Separation of hornblende and magnetite at this stage could
logically have played a role in altering the course of differentiation
to a trend marked by accelerated silica and alkali enrichment and
depletion of iron and magnesium in the succeeding differentiates,
icelandites and leucodiorites. Data on the amphibole from a typical
ferrogabbro, 67-J, are given on page 162 and in Table 17. Although
the amphibole has been converted to an oxyhornblende, its chemical
composition is probably close to that of the original form before
reheating oxidized much of the iron. It is low in silica compared to
the total rock or clinopyroxenes and correspondingly rich in alumina.
The effect of separation of such an amphibole would be to enrich the
residual liquid in silica at a greater rate than would be the case if
only pyroxenes were precipitated.

Iron oxide minerals are exceptionally abundant in ferrogabbros
(8.6 to 10.8 percent by volume), but subsequent rocks such as leucodiorites (67-D and 67-E) show a decrease in magnetite with little

or no decline in total iron (Tables 13, 14). Although MgO is depleted, the iron content remains nearly constant despite heavy precipitation of magnetite in more basic rocks.

The possibility of relating the ferrogabbros and leucodiorites to precipitation of the observed crystalline phases can be tested. The compositions of the amphibole and augite in a ferrogabbro (67-J) have been determined by analysis; the composition of the plagioclase in the same rock can be determined optically, and the compositions of iron oxides can be estimated from the material balance of iron and titanium in the total rock and the modal proportion of opaque minerals. These four minerals, which account for 96.8 percent of 67-J, can be asumed to have precipitated from a liquid that had somewhat higher Fe/Mg and Na/Ca ratios and a slightly higher SiO_2 content than the total composition of the rock in which they are now found. A suitable composition is that of the next more differentiated ferrogabbro, 67-I, which has a composition intermediate between that of the rock in which the minerals now occur (67-J) and the leucodiorite which is to be tested as a possible residual liquid (67-D). The compositions in terms of weight fractions of the major components are as shown in Table 19.

An equation can be written for the distribution of each component between these six phases. For example:

$$X_{SiO_2}^{fg} = X_{SiO_2}^{h} + X_{SiO_2}^{a} + X_{SiO_2}^{p} + X_{SiO_2}^{mt} + X_{SiO_2}^{ld}$$

where $X_{SiO_2}^{fg}$, $X_{SiO_2}^{h}$, and so on are the weight fraction of SiO_2 in the system that is found in the ferrogabbro, hornblende, and so on.

TABLE 19. Weight Fractions of Major Components in Two Differentiated Plutonic Rocks and in the Individual Mineral Phases by Which the Two Rocks Can Be Related through Crystal Fractionation

	Parent Liquid	Separated Crystals				Residual Liquid
	Ferrogabbro (67-I)	Hornblende (67-J)	Augite (67-J)	Plagio-clase (An_{50})	Ti and Fe Oxides	Leucodiorite (67-D)
SiO_2	0.502	0.434	0.511	..	..	0.554
Al_2O_3	0.164	0.099	0.013	0.284	..	0.150
FeO	0.107	0.150	0.145	..	0.755	0.114
MgO	0.036	0.111	0.114	..	..	0.019
CaO	0.088	0.120	0.200	0.104	..	0.051
$Na_2O + K_2O$	0.054	0.030	0.004	0.057	..	0.091

By solving six simultaneous equations, one for each component, we have calculated the proportion of each mineral that must be subtracted from the ferrogabbro liquid and the amount of leucodiorite liquid remaining. The weight proportions were then recalculated to volume percentages for comparison with the observed modal composition of 67-J.

The results show that the proportions of plagioclase, augite, hornblende, and iron oxides are close to those of the mode of 67-J.

	Calculated (%)	Mode (67-J)
Plagioclase	57.2	53.6
Augite	11.7	19.1
Hornblende	28.0	18.4
Iron oxides	3.1	8.9

The differences between the two proportions can be explained, at least in part, as the result of incomplete separation of crystals and the remaining liquid. The leucodiorite composition used in the calculation may not be that of the actual residual liquid because subsequent differentiation is known to have produced a quartz syenite. Nevertheless, the calculated proportion of residual liquid, 41.9 percent by volume, is large and shows that only a moderate amount of crystal fractionation is required to produce substantial changes in the trend of the liquid.

Presnall (1966) has shown that iron enrichment may continue despite early precipitation of magnetite if a magma crystallizes under conditions of low oxygen fugacity and constant total composition. On the other hand, Best and Mercy (1967) have shown that a differentiated calc-alkaline intrusion in California crystallized iron oxides under conditions of oxygen fugacity that were almost as low as those of the Skaergaard magma, and yet iron enrichment was negligible. They point out that some factor other than the fugacity of water and oxygen must play an important role in determining the course of differentiation.

Alteration of magnetite to maghemite and hematite in the Jervis Island rocks precludes the use of oxide minerals to deduce temperatures and oxidation conditions at the time of crystallization. The abundance of hornblende in many basic members of the series indicates that the magma crystallized under conditions of relatively high partial pressures of water. In later rocks hornblende declined in abundance and was replaced by ferropigeonite (which has partially inverted to ferrohypersthene). There is insufficient experimental data on amphiboles and iron-rich pyroxenes to interpret the possible significance of the pyroxene-amphibole transition; the complex interaction of changing iron-magnesium ratios, fugacity of water, silica concentration, and other factors combine to have a total effect on the crystallization of the ferromagnesium minerals that is unpredictable on the basis of present knowledge.

SIMILARITIES TO OTHER VOLCANIC SERIES

The Galápagos lavas and those of the Brito-Icelandic province show some remarkable similarities, despite the differences in their tectonic settings. The tholeiitic, alkali-olivine, and feldspar-phyric basalts of the Galápagos are chemically and mineralogically equivalent to the nonporphyritic central, plateau, and porphyritic central types of Mull. The reversed titania relation of the Galápagos basalts, in which the highest contents of titania are found in iron-rich tholeiites, does not seem to have been observed in analogous Hebridean basalts, but this may be because of less accurate determinations of titania in the older analyses. More recent analyses of Icelandic lavas by Carmichael (1964) show as much as 3.6 percent titania in iron-rich Tertiary tholeiites of Thingmuli Volcano. Iron-rich tholeiites of the Faeroes are also characterized by high titania contents (Walker and Davidson, 1936; Noe-Nygaard, 1967.)

There are also many similarities in composition, as well as in temporal and spatial relations, between the Galápagos and the Brito-Icelandic differentiated suites. Especially pronounced is the similarity between the intermediate and acid rocks of Jervis and Duncan islands on the one hand and those of the normal magma series of Mull on the other (Bailey and others, 1924, Fig. 2 and Table III). Few Galápagos rocks correspond to the Mull alkaline magma series of mugearites and trachytes, but future workers may find rocks of this type among the products of late parasitic vents on Indefatigable Island and the western part of Chatham Island.

Eastern Pacific Islands

The islands nearest the Galápagos archipelago, Cocos and Malpelo, are small and neither has been thoroughly studied. Both are entirely volcanic, but whereas Cocos Island is distinctly oceanic in character, Malpelo may have some relation to the adjacent continent.

The rocks of Cocos Island, collected for us by Dr. Allan Cox, are mainly alkali-olivine basalts with abundant phenocrysts of olivine and labradorite and a few of titaniferous augite in an intergranular groundmass of labradorite, olivine, clinopyroxene, and ore. Among the differentiated rocks are coarse-grained trachytes containing oligoclase, anorthoclase, ferroaugite, iron-rich olivine, ore, and unusually large and abundant prisms of smoky apatite. An analysis of one such specimen is presented in Table 20 (No. 3).

Rocks collected from Malpelo Island during a brief examination by members of the Papagayo Expedition of Scripps Institution of Oceanography are quite different from those of Cocos Island. They include a large proportion of very glassy basalts, undoubtedly of submarine origin, but there are also quartz diabases, possibly intruded as sills within the uplifted submarine sequence. A typical diabase is made up of randomly oriented laths of labradorite, ophitic augite, chlorite, "chlorophaeite," ore, calcite, and about 5 percent late-

TABLE 20. NEW ANALYSES OF ROCKS FROM LOCALITIES IN THE EASTERN PACIFIC OTHER THAN THE GALAPAGOS ISLANDS

	1	2	3	4	5
SiO_2	49.58	47.88	58.07	46.82	63.61
TiO_2	2.24	0.88	1.63	3.84	0.33
Al_2O_3	14.47	20.33	18.84	16.50	18.89
Fe_2O_3	1.94	1.74	3.27	2.83	1.08
FeO	10.27	5.29	2.05	8.45	1.39
MnO	0.25	0.13	0.10	0.17	0.12
MgO	6.52	7.59	0.99	6.12	1.20
CaO	10.81	13.10	3.12	8.14	1.59
Na_2O	2.62	2.05	5.34	3.60	7.99
K_2O	0.25	0.14	4.31	1.54	3.35
H_2O+	0.39	0.55	1.12	0.58	0.26
H_2O-	0.06	0.06	0.28	0.42	0.25
P_2O_5	0.20	0.09	0.64	0.79	0.12
Total	99.60	99.83	99.76	99.80	100.18
Ap	0.42	0.18	1.25	1.68	0.24
Il	3.18	1.22	2.30	5.48	0.44
Or	1.50	0.85	25.75	9.40	19.15
Ab	24.05	18.40	48.40	30.07	69.35
An	27.45	45.72	11.90	25.02	5.62
C	..	..	1.13	..	..
Mt	2.07	1.80	1.35	3.03	1.09
Hm	..	..	1.39	..	..
Di	20.89	15.04	..	9.08	1.04
Hy	20.14	8.66	2.95	..	0.88
Ol	..	8.13	..	14.39	2.19
Q	0.30	..	3.58	..	..
Ne	..	..	..	1.85	..
Total	100.00	100.00	100.00	100.00	100.00
si	118.1	107.4	206.8	110.5	270.0
al	20.2	26.7	39.4	23.5	34.5
fm	45.8	37.1	20.4	44.7	16.4
c	27.6	31.5	11.9	21.1	7.2
alk	6.4	4.7	28.2	10.8	41.8
k	0.059	0.043	0.347	0.220	0.217
mg	0.507	0.687	0.259	0.497	0.465
qz	—7.5	—11.4	—6.0	—32.7	+2.8

1. Basaltic glass, East Pacific Rise, 12°52′S., 110°57′W., amphitrite dredge haul 3.
2. Porphyritic feldspar-rich basalt, same dredge haul as 1.
3. Trachyte, Cocos Island, specimen collected by Allan Cox.
4. Alkali basalt, Masatierra Island, collected by M. N. Bass.
5. Sodalite trachyte, Masatierra Island, collected by M. N. Bass.

crystallizing quartz. The texture and composition of this rock resembles those of diabase sills within the Nicoya Complex of Costa Rica. Chesterman (1963) described a pyroxene andesite from Malpelo Island, collected by the California Academy of Science Expedition of 1906, but none is present in our collection.

The closest analogues to the Galápagos differentiated suite are found on San Benedicto and Easter islands. Although no Galápagos rocks are as siliceous as the soda rhyolite of Easter Island, other members of the two differentiated suites are remarkably similar in their over-all chemical composition. Enrichment in iron-magnesium ratios is even more pronounced in the Easter Island "andesites" than it is in lavas of comparable silica content from Jervis and Duncan islands. Pigeonite-bearing types are present on Easter Island and in the Galápagos group, and the basalts of Easter Island include both tholeiitic and alkaline types; large phenocrysts of plagioclase seem to be common. As already noted, the rates of alkali and silica enrichment in the two suites are very similar (Fig. 41).

Richards (1966) has described the rocks of San Benedicto as a differentiated alkaline suite, mainly on the basis of alkali-silica ratios. Unfortunately, there are few basalts exposed on the island

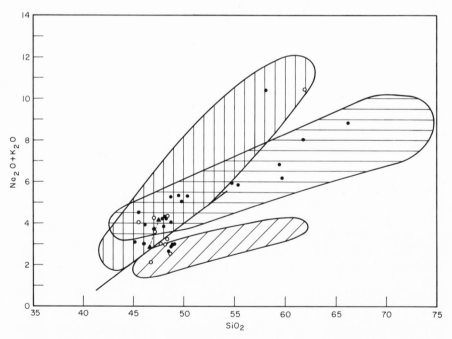

Figure 41. Silica-alkali relations of the rocks of the Galápagos, Easter, and Hawaiian islands. Vertically and obliquely ruled areas enclose the Hawaiian alkaline and tholeiitic rocks, respectively. Horizontally ruled area encloses the Easter Island rocks. New analyses of Galápagos rocks are shown as closed circles, old analyses by open circles.

and it is impossible to know the composition of lavas in the submarine portion of the volcanic pile. The differentiated rocks show a progressive enrichment in silica with increasing amounts of normative quartz, which reaches 18.9 percent in a sodic rhyolite. This feature, as well as other characteristics of the differentiated series, suggests a closer affinity to tholeiitic rocks, such as those of Easter Island and the Galápagos. The problem of interpreting such trends in terms of alkali-silica ratios alone has already been pointed out (p. 153).

There are too few analyses of other islands in the eastern Pacific to make close comparisons. Judging from fragmentary data, it appears that the differentiated rocks of this region share several chemical features, such as a dominance of sodium over potassium and a tendency for rather high content of phosphorus. Some rocks are decidedly pantelleritic. But most important, they show considerable variation in degree of silica saturation. The most siliceous rocks, with the highest percentages of normative quartz and Niggli qz values, seem to be those associated with the East Pacific Rise. Undersaturated trachytes and phonolites are found only on islands far from the crest of the rise. These relations are discussed at greater length below.

Submarine Lavas

Data on submarine lavas from the floor of the eastern Pacific are limited to a small number of widely scattered dredge hauls, mostly from the East Pacific Rise. Careful studies of the submarine lavas, comparable to those of Muir and Tilley (1964, 1966) on the Mid-Atlantic Ridge, have yet to be carried out in the Pacific.

The nearest dredge hauls from which we have been able to obtain samples lie many miles from the Galápagos Platform. Professor G. Arrhenius supplied samples from two points on the East Pacific Rise, about 1100 and 1400 miles southwest of the archipelago, and Professor M. N. Bass obtained excellent samples from the escarpment of the Clipperton Fracture Zone about 800 miles north of the Galápagos Islands, as well as others from near Malpelo Island, about 400 miles to the east. In addition, other members of the Scripps Institution of Oceanography have permitted us to examine material dredged from numerous seamounts throughout the central and eastern Pacific. We have no way of knowing to what extent these specimens are representative of the lavas of this vast region, but certain chemical and petrographic features that the rocks share with the Galápagos lavas suggest that they are quite similar and may be typical of a very wide area.

The submarine lavas that we examined differ markedly in texture from those of the emergent volcanoes. Most of them have well-developed variolitic textures, with radiating intergrowths of labradorite laths and subcalcic augite in a matrix of clear brown glass interlaced with chains of magnetite and ilmenite. Olivine is rare or totally absent, and phenocrysts of clinopyroxene appear to be lacking. Four specimens from the Clipperton Fracture Zone vary only in the proportion of glass to crystalline phases and they may well have come from the same lava flow. One specimen from the East Pacific Rise (No. 1, Table 20) consists almost entirely of tachylite with a shell of clear sideromelane. The tachylite is made up of small ovoid masses of faintly birefringent and nearly opaque material. The only mineral detectable in an X-ray diffraction analysis is a trace of clinopyroxene. Another specimen (No. 2, Table 20) contains euhedral phenocrysts of labradorite-bytownite, up to 1 cm across, and rare microphenocrysts of magnesian olivine in a fine-grained variolitic groundmass of elongated augite crystals, labradorite, ore, and opaque glass. The chemical composition of the aphyric glass is almost identical to that of Galápagos tholeiitic basalts, although, like that of most submarine lavas, the oxidation state of the iron is somewhat lower than that of subaerial lavas. The feldspar-phyric basalt differs from that of the strongly porphyritic Galápagos basalts in containing normative hypersthene rather than nepheline, and it has a somewhat lower iron-magnesium ratio.

Siliceous volcanic glass has been discovered throughout a wide area of the eastern Pacific. The extensive Worzel ash (Ewing and others, 1959) seems to have been derived from large pumice eruptions on land; but quartz, alkali feldspars, and glass of low refractive index (described by Peterson and Goldberg, 1962) seem to be closely associated with the crest of the East Pacific Rise and are believed to be of submarine origin. Unfortunately, no chemical data are available on the glass, but the sodic nature of the alkali feldspars is consistent with the proposed derivation of the siliceous material from submarine eruptions.

PETROGENETIC RELATIONS TO OTHER VOLCANIC ROCKS OF THE PACIFIC

Opinions on the characteristic composition of oceanic rocks have passed through rapid and extreme oscillations within a few decades. Until quite recently, it was the consensus of petrologists that oceanic basalts are entirely of the alkali-olivine type, high in magnesium, titanium, and alkalis and low in silica. Even though Lacroix, as long ago as 1928, pointed out the wide range of silica saturation of igneous series in the Pacific, it was not until the significance of tholeiites in Hawaii was recognized that the importance of silica-

saturated basalts was generally accepted. When dredge hauls began
to bring fresh samples from the deep ocean floors and most of these
showed tholeiitic affinities, the pendulum of opinion reached the
opposite extreme and it was announced that tholeiites were in fact
the only true "primitive" rocks, all others being products of differentiation of a single primary magma of very uniform composition.
Despite the fact that only a handful of specimens has been studied
and most of these came from the more accessible regions of the mid-
oceanic ridges, compositions of "average oceanic tholeiites" have
been defined with the specification that they represent the purest,
indeed the only true, mantle-derived primitive magma.

Magmatic provinces have been recognized as characteristic features of continents since they were first pointed out by Volgelsang
in 1872 and Judd in 1886. Interpretation of regional and temporal
variations in igneous rocks has since become one of the most
important aspects of igneous petrology. It is only reasonable, therefore, that one would expect variations in noncontinental regions and
that these should be related to large-scale tectonic features that have
played an important role in the evolution of the ocean basins. One
need only look at the striking contrasts between the rocks of such
regions as the central and eastern Pacific or the islands of the
central Atlantic and those along its margins to see that there are
regional differences, at least in silica saturation, on a scale too large
to be explained by accidents of fractionation.

The small area of the oceans occupied by emergent volcanoes
accessible to close study at first appears to be an insurmountable
obstacle to any attempt to define regional variations. In fact, however, two fortunate circumstances operate to make the islands
exceptionally valuable. First is the tendency of igneous magmas to
differentiate toward end products that magnify chemical characteristics only subtly revealed in the parent basalts. Second is the common
emplacement of these end products as resistant plugs that rise into
the volcanic conduits toward the close of activity and withstand the
onslaught of wave action long after all else is eroded to sea level.

Lacroix (1928) was one of the first petrologists to consider the
distribution of various kinds of volcanic rocks within the Pacific
Basin in terms of their chemical characteristics, and certainly the
first to stress the importance of relative silica and alkali contents
of oceanic basalts and their differentiates. The α basalts described
by Lacroix are oversaturated in silica and contain modal or normative silica minerals; his β group consists of an intermediate saturated
series without quartz or nepheline and a nepheline series containing
neither quartz nor hypersthene as modal or normative minerals.

Lacroix suggested that the Pacific volcanoes show regional
variations of both their parent basalts and their differentiates, and

he believed that these could best be defined by their degree of silica saturation. He pointed out that in some islands all the rocks are saturated regardless of their degree of differentiation, whereas on others the entire series may be strongly undersaturated. The most common relation, however, was thought to be that in which trachytes with no normative or modal nepheline are associated with β basalts and mugearites.

More recent data indicate that Lacroix's scheme is valid, offering an excellent means whereby volcanic islands can be considered from a petrological standpoint. Our joint investigation of the Galápagos rocks, as well as our independent studies of the rocks of Tahiti (Williams, 1933; McBirney and Aoki, 1968), lead us to conclude that the three major petrologic groups outlined by Lacroix are well represented by Tahiti, Hawaii, and the Galápagos. The lavas and equivalent plutonic rocks of Tahiti form an outstanding example of a well-differentiated and strongly undersaturated series that can be contrasted with the tholeiitic series of the Galápagos Islands; the lavas of Hawaii display some of the features of both other regions, but in certain important respects they have no equivalents elsewhere in the oceans.

The Hawaiian Magmatic Suite

The temporal and geologic relations of the rocks of Hawaii are well-known. Tholeiitic basalts of nearly uniform composition constitute the vast bulk of the volcanoes, and only in late stages of declining activity do more varied rock types appear. Two subsequent trends then emerge. Basalts become more alkaline and increasingly undersaturated with silica as olivine tholeiites give way first to alkali-olivine basalts, then to nepheline basanites, and finally to nepheline-melilite basanites. Distinct from this basic series is the sequence of progressively more felsic hawaiites, mugearites, and trachytes, which have been interpreted as differentiates of a parental alkali-basalt magma. Rare "rhyodacites" and quartz dolerites are believed to be the corresponding differentiates of a tholeiitic parent.

These trends are clearly distinguished on a Von Wolff diagram, such as that used by Macdonald and Katsura (1962). Figure 42 shows the progression from mafic to felsic compositions during the normal course of differentiation of the Hawaiian alkaline suite as well as the relation of more siliceous differentiates of the tholeiitic magma.

The basaltic series, however, follows a trend almost normal to that of differentiated rocks. The differentiated rocks become increasingly felsic but show only a minor variation in silica saturation; the basalts become more mafic and progressively poorer in silica.

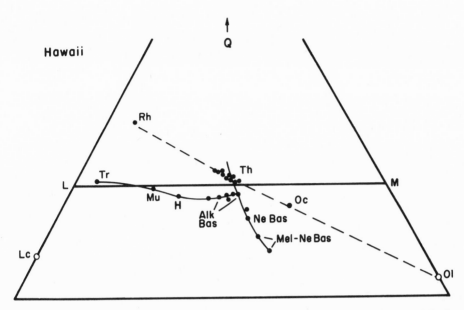

Figure 42. Von Wolff diagram for rocks of the Hawaiian Islands. The method of plotting is modified from the conventional one by including normative apatite with the leucocratic constituents (L) and normative magnetite and ilmenite with the mafic constituents (M). This method produces smoother variation curves and allows inclusion of all the normative components of a standard analysis. Two trends are clearly distinguished. The main differentiation series of hawaiite (H), mugearite (M), and trachyte (T) is believed to be derived from a parent alkali basalt. Less common siliceous differentiates, such as the "rhyodacite" (Rh), are believed to be derived from a tholeiitic parent, mainly by subtraction of olivine. The trend of compositions of basalts is distinct from that of differentiates. It proceeds from tholeiites (Th) through alkali basalts (Alk Bas) and nepheline basalts (Ne Bas) to melilite-nepheline basanites (Mel-Ne-Bas). Points are plotted from analyses given by Macdonald and Katsura (1962) and Winchell (1947).

The Galápagos Magmatic Suite

When the Galápagos rocks are plotted in the same manner (Fig. 43), the trends of differentiated rocks are again distinct from that of basalts. The effusive and plutonic series follow divergent differentiation trends differing in the degree of silica saturation of increasingly leucocratic rocks. This is in contrast to the principal types of basalts, tholeiites, olivine tholeiites, and alkali-olivine basalts, which vary along a trend normal to that of the differentiated rocks. The Galápagos series differs from the Hawaiian suite in that the principal differentiates are tholeiitic and are uniformly more siliceous than the hawaiite-mugearite-trachyte series of Hawaii. The more aluminous character of the basalts shifts them farther from the mafic side to a more intermediate position.

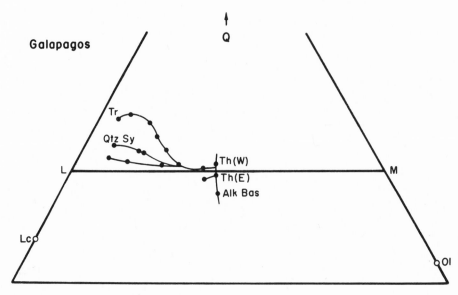

Figure 43. Von Wolff diagram for the rocks of the Galápagos Islands. Three principal types of basalt — tholeiites, Th (W), olivine tholeiites, Th (E), and alkali-olivine basalts, Alk Bas — fall along a trend that reflects their differing degrees of silica saturation. Tholeiites of the central islands differentiated to siliceous trachyte and quartz syenite and three other analyzed lavas appear to fall along an independent trend that is only slightly oversaturated in silica.

The Tahitian Magmatic Suite

The rocks of Tahiti are illustrated in Figure 44. One series is strongly undersaturated throughout and includes nepheline syenites and phonolites; the other contains little or no modal or normative nepheline and trends toward trachytes or syenites almost exactly saturated with silica.

The most common basalts of Tahiti are moderately undersaturated ankaramites. They are chemically similar to the nepheline basanites of Hawaii but commonly contain abundant phenocrysts of olivine and clinopyroxene. An aphyric alkali basalt has been considered a possible parent of the differentiated suite (McBirney and Aoki, 1968, p. 546-547), but this interpretation is probably incorrect. O'Hara (1965, 1968) has pointed out that lavas that are erupted at the surface free of phenocrysts but on or close to their liquidus temperature cannot represent a magma that has risen unchanged from greater depths. It is more likely, therefore, that the bulk composition of porphyritic lavas is closer to an original composition than that of aphyric basalts. The Tahitian ankaramites are more mafic than the previously proposed parental magma but lie on the same trend shown in Figure 44. A greater range of basic rocks is

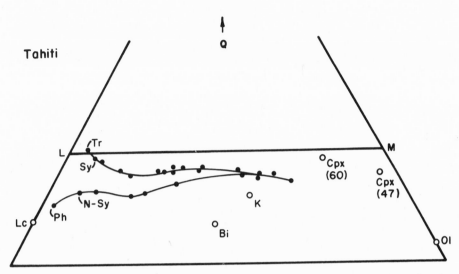

Figure 44. Von Wolff diagram for the rocks of Tahiti. Two differentiated series are shown, one with modal feldspathoids trending toward phonolite and nepheline syenite, the other toward trachyte and syenite. Compositions of analyzed biotite, kaersutite, and clinopyroxenes from basic plutonic members of the suite are also shown.

needed before it will be possible to say whether there are compositional variations among the basalts other than the main differentiation trends.

REGIONAL RELATIONS OF VOLCANIC ROCKS OF THE PACIFIC

The variations of silica saturation observed in the volcanic islands of the eastern Pacific (Fig. 45; Table 21) appear to be related to regional differences of heat flow (McBirney and Gass, 1967). Tholeiitic basalts and markedly siliceous differentiates, such as those of the Galápagos and Easter islands, are confined to a narrow region near the crest of the East Pacific Rise where heat-flow values are abnormally high (Fig. 46). Strongly undersaturated basanites and phonolites, such as those of Tahiti and San Felix, are found in regions of low surface heat flow far from the axis of the rise. A comparison of heat-flow measurements and the degree of silica saturation of islands in the eastern Pacific is shown in Figure 47. A similar distribution is observed in the Atlantic Ocean. These relations have been interpreted as a reflection of the depth of magma generation and differentiation, low heat flow indicating a low thermal gradient that would reach melting temperatures at a greater depth than the steep gradient required for high heat flow.

Such an explanation of the compositional differences of volcanic rocks is consistent with several lines of experimental evidence that show that liquids produced by partial melting of mafic or ultramafic rocks become increasingly undersaturated with silica at higher pressures. Kushiro (1965), for example, has shown that the invariant points in the systems forsterite-$CaAl_2SiO_6$-silica and forsterite-nepheline-silica migrate toward increasingly silica-deficient compositions as pressure increases. Figure 48 (*after* Kushiro's Fig. 22), showing this shift toward silica-poor liquids, is a close analogue of the Von Wolff diagrams (Figs. 42-44) used to show the trends of composition found in Hawaii, the Galápagos, and Tahiti, because anorthite corresponds to the feldspar component (L), enstatite to the saturated mafic minerals (M), silica to normative quartz, and $CaAl_2SiO_6$ and olivine to the undersaturated components. It has been suggested (McBirney, 1967) that the trend of compositional variations in oceanic basalts can be directly correlated with the pressure

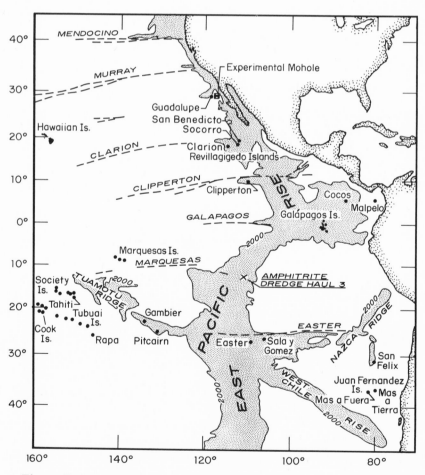

Figure 45. Generalized map of the eastern Pacific Ocean showing the locations of the principal islands and structural features.

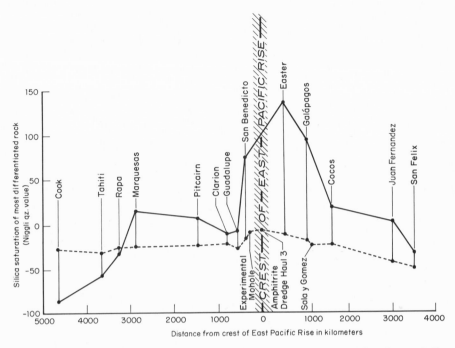

Figure 46. Relations of the degree of silica saturation of basalts and the most differentiated rocks of volcanic islands to the position of the islands with respect to the East Pacific Rise. (*From* McBirney and Gass, in 1967.)

relations of Kushiro's study and hence with the depth from which individual magmas were derived.

The Dual Nature of Oceanic Tholeiites

When one attempts to interpret the Hawaiian suites in a similar fashion certain inconsistencies arise. We have already pointed out some of the differences between the hypersthene-normative rocks of Hawaii and those of the Galápagos and other islands near oceanic rises (p. 172). Most conspicuous is the fact that the Hawaiian tholeiites contain olivine phenocrysts but few of plagioclase while the reverse is true in the eastern Pacific and mid-Atlantic regions. Hawaiian tholeiites are richer in MgO, Ni, and Cr and poorer in TiO_2 and CaO than are their associated alkali basalts, again the reverse of relations elsewhere in the oceans. The most strongly differentiated suite in Hawaii is derived from an alkaline magma, but in the Galápagos, Iceland, Easter, and other islands near the oceanic rises it is the tholeiitic suite that is most complete and the alkaline rocks, if present, show little tendency toward differentiation. The only important oceanic suite we know of that closely resembles that of Hawaii is the one found on Réunion (Upton and Wadsworth,

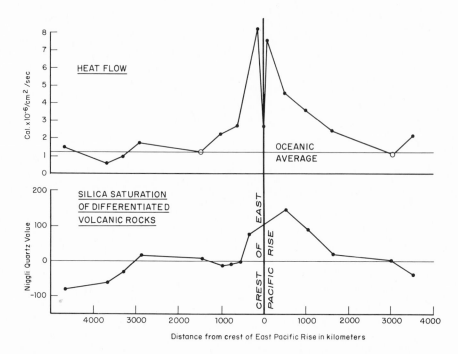

Distance from crest of East Pacific Rise in kilometers

Figure 47. Comparisons of heat-flow profiles and silica-saturation curves of differentiated rocks across the East Pacific Rise. Heat-flow profile is compiled from data of Von Herzen (1959), Von Herzen and Uyeda (1963), and Langseth and others (1965) using heat-flow measurements nearest to individual islands. Solid points on the Pacific profile are from heat-flow measurements within 5° of an island; open circles are for measurements more than 5° but less than 10° from the islands. The geological relevance of individual points on the heat-flow profile varies greatly from island to island, and the curve cannot be interpreted as more than a qualitative representation of the heat flow in the vicinity of the islands.

1965). There, as in Hawaii, the main masses of the volcanoes are built of olivine tholeiites, and these are followed by ankaramites and a differentiation series that includes hawaiite, mugearite and trachyte.

Green and Ringwood (1967) propose that Hawaiian olivine tholeiites are products of partial melting of a rising diapir of "pyrolite". They assume a starting composition of 25 percent olivine tholeiite and 75 percent peridotite and then deduce that the melting mechanism under Hawaii results in 25 percent olivine tholeiite and 75 percent residuum. O'Hara (1968, p. 93) has pointed out some of the inconsistencies of this model, the most serious of which is that the experimental work of Green and Ringwood showed that olivine is not a liquidus phase for their olivine tholeiite at pressures above 9 kilobars. Hence, such a composition cannot be a partial melt of olivine-bearing peridotite or "pyrolite" at such pressures, nor can it be the

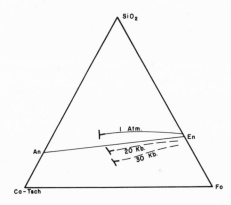

Figure 48. Diagram showing migration of the invariant point in the system $CaAl_2SiO_6$ (Ca-tschermakite)—forsterite—silica with increasing pressure, according to Kushiro (1965a). The composition shown by the heavy intersecting lines corresponds to that of the first liquid produced by partial melting at the indicated pressure.

product of fractionation of olivine from another liquid from greater depth. One possible explanation for this anomaly is that the olivine tholeiite is a derivitive liquid produced at a depth where olivine has been precipitated and removed from the magma before it reached a reaction relation to clinopyroxene, as suggested by the experimental results of Green and Ringwood (1967, p. 143).

O'Hara (1965, 1968) has offered an alternative explanation for olivine tholeiites. He points out that the composition of the first liquid produced by partial melting of an assemblage of olivine, orthopyroxene, clinopyroxene and an aluminous phase (plagioclase, spinel, or garnet, depending on pressure) does not continue to become more and more undersaturated with silica at increasing depths but reaches a limit and may even reverse this trend. Because of the large expansion of the primary field of garnet at pressures above about 20 kilobars, he deduces that the first liquid may even become hypersthene-normative again, so that the depth interval in which nepheline-normative liquids may be produced could be bounded both above and below by regions in which hypersthene-normative liquids are in equilibrium with mantle assemblages. The precise pressures and depth levels of these boundaries are very sensitive to the proportion of Na to Ca in the system and are difficult to evaluate.

This interpretation would lead one to infer the possibility of two distinct oceanic tholeiites derived from quite different levels. The olivine tholeiites of Hawaii could represent magma produced below the level of nepheline-normative liquids and the plagioclase-rich tholeiites of the eastern Pacific and mid-Atlantic Ridge would originate at much shallower depth. Alkali basalts may be produced in the lower part of a zone of magma generation in regions of shallow melting but come from the upper part of such a zone under Hawaii. The relative degrees of differentiation in the two situations would be consistent with the higher-level magma, being in both cases that which is most strongly differentiated.

Thermal Aspects of the Generation and Rise of Magmas

These speculations on the compositional relations of basalts and their differentiates are inherently dependent on the mechanism by which the liquids are produced and rise to the surface. Although there are now many experimental data on crystal-liquid relations at high pressures, it is difficult to relate this information to the compositions of lavas erupted at the surface. As O'Hara (1965) has pointed out, the fact that aphyric lavas are erupted at temperatures close to a three- or four-phase liquidus at one atmosphere demonstrates that these same liquids could not have been in equilibrium with mantle source-rocks at high pressure. Porphyritic rocks, such as picrites, feldspar-phyric basalts, and ankaramites, are probably closer to the bulk composition of the original liquid, but there remains the question of what portion of the phenocrysts was derived from the host liquid and what the effects of wall-rock reactions may have been as the liquid rose through the mantle and crust. This is essentially a thermo-dynamic problem, not easily assessed quantitatively.

Most proposed mechanisms for generation of oceanic magmas call on increased heat flux or a decrease of pressure to produce partial melting of the mantle. Quantitative evaluations of these mechanisms depend on a knowledge of the thermal and mechanical properties of rocks at high temperatures and pressures, and in many respects this knowledge is regrettably inadequate. It is possible, however, to place qualitative limits on these factors and examine their role in various models for magma generation.

Stage 1 — The initial heating cycle. The most serious restraint on mechanisms of partial melting, whether due to a temperature rise or a pressure drop, is the one imposed by the relatively low heat capacity and high heat of fusion of rocks. The fact that the specific heat of most igneous rocks (about .3 cal/gm) is more than two orders of magnitude lower than the heat of fusion (about 100 cal/gm) limits sharply the amount of melting that can take place without wholesale heating of very large volumes of rock to their melting temperature. Hess (1960), for example, has shown that melting caused by a localized temperature anomaly in a layered mantle would proceed very slowly because of the rapid lateral heat dissipation below the horizon where melting takes place. Vertical temperature gradients are probably more pronounced in the mantle than are horizontal ones; so in the general case of an unlayered mantle, heat will be dissipated vertically even more rapidly than laterally.

The effect of volumetric expansion on melting (about 10 percent), together with the low tensile strength of rocks, even under high confining pressures, imposes an additional mechanical limit on the amount of liquid that can be produced. Bott (1965) and Robson and others (1968) have assessed this problem and conclude that only about 0.5 percent of the rock could melt before stresses due to volumetric expansion would cause the rock to fail. The liquid, filter-pressed from its host, moves upward in the direction of the least principal stress and passes into rocks at progressively lower temperature. Owing to the small volume of the liquid and the narrow dimensions of the fractures through which it must move, the heat content of the liquid will be quickly transferred to the walls of the fracture and augment the tendency for rocks above the zone of melting to be raised to their melting temperature.

It is helpful to examine some of the quantitative aspects of this condition. We can estimate the distance, z, that a liquid will move upward before it crystallizes by equating the heat, Q_m, carried by the magma per horizontal unit length of the channel to the heat flux, Q_k, conducted through a corresponding area of the walls.

$$Q_m = Q_k \tag{1}$$

Q_m can be taken as the product of the mass of liquid moving up the fracture per unit of time and the sum of its heat capacity and heat of fusion.

$$Q_m = V\rho h \left[C_p \frac{z}{2} \left(\frac{dT_m}{dz} \quad \frac{dT_a}{dz} \right) + \triangle H_f \right], \tag{2}$$

where V is velocity, ρ density, h width of the fracture, C_p heat capacity, $\triangle H_f$ the heat of fusion, and $\frac{dT_m}{dz}$ and $\frac{dt_a}{dz}$ the melting temperature and adiabatic gradients, respectively. The velocity, V, can be estimated from the equation for viscous flow between parallel walls (Kay, 1963, p. 155).

$$V = \frac{\triangle P h^2}{12\eta z}, \tag{3}$$

where $\triangle P$ is the pressure difference between the ends of the channel and η is the viscosity. $\triangle P$ cannot be greater than the hydrostatic difference of the weight of the columns of liquid and solid rock, $\triangle \rho z$, plus the pressure, $P\sigma$, released when the rock fails as a result of expansion on melting.

Assuming the wall rocks 1 cm from the contact are at a temperature governed by a linear conductivity gradient, $\frac{kdT}{dz}$, and that the gradient out from the contact is also linear, the heat flux through the two walls is approximately

$$Q_k = 2k \left(\frac{dT_k}{dz} - \frac{dT_m}{dz} - \frac{dT_a}{dz}\right) \frac{z^2}{2} , \qquad (4)$$

where k is thermal conductivity.

Equating these two equations for the heat flux of the magma and the heat conducted through the walls of the channel, substituting the following values:

$$P\sigma = 200 \text{ bars}$$
$$\rho = 2.5$$
$$\triangle\rho = 0.5$$
$$\eta = 103 \text{ poises}$$
$$C_p = .3 \text{ cal gm}^{-1}$$
$$\frac{dT_m}{dz} = 3° \text{ km}^{-1}$$
$$\frac{dT_a}{dz} = .3° \text{ km}^{-1}$$
$$\frac{dT_k}{dz} = 10° \text{km}^{-1}$$
$$\triangle H_f = 100 \text{ cal gm}^{-1}$$
$$k = 4{\cdot}10^{-3} \text{ cal cm}^{-2} \text{ sec}^{-1}$$

and simplifying, we obtain

$$10.5{\cdot}10^{-4} z^3 - 0.2{\cdot}10^{-5}h^3z^2 - 50h^3z - .81h^3z - 2{\cdot}10^7h^3 = 0 . \qquad (5)$$

For values of h less than 10 m all but the first and last terms are negligible, so that the equation becomes essentially

$$z = 2.7{\cdot}10^3 h . \qquad (6)$$

Thus, for a fracture with width h equal to 1 cm, magma can rise about 27 m before it loses enough heat to crystallize completely, if the overlying rocks lie on a conductivity gradient. In more restricted channels, the distance is correspondingly less.

It is apparent that magma produced at depths in the mantle where upward passage is restricted to sinuous fractures cannot reach the surface as long as the wall rocks above the level of melting are significantly below their melting temperature. Instead, as heat enters the system and partial melting proceeds, a front of liquid will advance upward while shallower rocks are raised to their melting temperature. Only when a large vertical column of rock is on the melting curve up to a depth at which fractures several tens of centimeters wide are possible can magma rise to the surface.

Stage 2 — The main effusive cycle. Once mantle rocks are on the melting curve, the liquid will no longer dissipate its heat of fusion, but it will still add to the walls the potential superheat which results from lowering of the melting temperature with decreasing pressure. A slight amount of cooling accompanies the adiabatic expansion of the liquid, but this is substantially less than the melting temperature gradient (about 0.3° vs. 3° per kilometer), so that in the absence of conductive heat losses the liquid would gain about .81 cal gm^{-1} of superheat for each kilometer it rises. That superheated lavas are not observed at the surface testifies to the fact that there must be conductive heat losses to the walls of the channel, and we have already seen from the foregoing calculation that this heat transfer must be efficient. It must be concluded, therefore, that the potential superheat is transferred to the wall rocks as the magma rises toward the surface.

The primary field of olivine increases steadily as basaltic liquids ascend to lower pressures, and olivine must be precipitated if the temperature descends along the olivine-pyroxene liquidus. This will be true as long as pyroxene is present in the walls. The amount of olivine crystallizing is shown by the data of Kushiro (1968) to be about 5 weight percent between 20 and 30 kilobars or approximately .2 percent per kilometer. Crystallization of this amount of olivine will add about .22 cal gm^{-1} km^{-1} to the amount of available heat. Thus a total of about 1.03 cal gm^{-1} will be lost by the magma and gained by the wall rocks for each kilometer the liquid rises.

This heat cannot cause simple melting but will be consumed in reaction between the magma and wall rocks. Owing to the fact that the liquid is already saturated with olivine, the phases dissolved from the walls will be mainly pyroxenes, of which enstatite will be the most important. For each kilometer of rise through heat-saturated horizons, the liquid crystallizes .2 weight percent olivine and can dissolve about .7 percent enstatite. If the liquid is initially alkaline and the superheat is great enough to surmount a low thermal barrier, the liquid can be reduced in its normative nepheline content at the rate of .35 percent for each kilometer. Thus, in a distance of only 10 kilometers, a strongly alkaline liquid can be converted to one which is hypersthene-normative, namely an olivine tholeiite. During early stages of reaction, a combination of both enstatite and clinopyroxene will be dissolved, and if the clinopyroxene is aluminous, the degree of silica enrichment of the liquid will be slightly reduced, but this effect will be quite subordinate.

These considerations indicate that the compositions of magmas rising from considerable depths in the mantle will be significantly alerted by wall reaction, the principal effect of which will be to impart to the magmas a tholeiitic character. Such liquids will be distinguishable from hypersthene-normative basalts derived from shallow depths by their high olivine and low plagioclase content. The mechanism provides a possible explanation for the marked contrast between the olivine tholeiites of Hawaii and the plagioclase tholeiites of oceanic ridges.

Stage 3 — Declining activity and differentiation. Depending on the bulk composition of the mantle, successive minerals will be depleted by prolonged reaction with rising magmas, and olivine will eventually be the last remaining crystalline phase. When the wall rocks have been entirely converted to dunite, the heat they absorb from the rising magma is no longer consumed in melting pyroxene and can raise the temperature of the wall rocks above the temperature of the olivine-pyroxene liquidus. From this point on, the temperature will steadily move up an olivine liquidus as the liquid becomes progressively more picritic without becoming more siliceous.

A second effect becomes important as eruptions become less frequent and the rate of heat transfer by the liquid declines. Wall rocks will begin to fall off from their liquidus temperature at progressively deeper levels, and magma will rise through greater intervals in which the temperature of the walls is below the melting curve. As long as the walls remain below the liquidus temperature of the magma further melting or reaction is impossible; heat given up by the magma is dissipated by conduction rather than by melting. More important, the temperature of the magma can also fall below the two-phase olivine liquidus and other minerals such as pyroxenes will begin to crystallize.

These conditions are ideal for differentiation of the magma as it rises toward the surface. As crystals precipitate on the walls and grow from the liquid passing over them, a highly efficient mechanism of crystal fractionation is established. The degree of differentiation will depend, of course, on many factors, such as the interval of time between eruptions and the rate and volume of flow. The compositions of differentiated rocks erupted in these declining stages are more likely to show irregular variations between a basic parent and felsic end-product than to constitute a regular sequence of progressively more felsic rocks. This is in accord with the wide compositional variations and frequent reversals observed in the late lavas of many volcanic centers.

The nature of the trend of the differentiation series depends on at least two factors—the composition of the magma when it reaches the level at which wall rocks fall below the melting curve and the depth at which this transition is reached. Green and Ringwood (1967) and others have determined the mineral phases precipitated from various basaltic compositions at different temperatures and pressures, but in order to relate these data to differentiation of a rising magma column it is necessary to integrate the effects throughout a wide depth range. Without a better understanding of the rates of rise and heat loss it is difficult to make any meaningful calculation of the end product of fractionation of a given liquid.

In view of the complexity of the mechanism, it is remarkable that differentiation commonly follows a consistent trend or pattern throughout the visible history of volcanism. Most commonly, a single differentiation series is dominant, and the nature of this trend reflects the composition of the magma from the shallowest part of the zone of magma generation. On many islands, such as Tahiti and Kerguelen, differentiates follow two distinct courses, with contrasting series appearing side by side during a period of declining activity.

The divergent trends of many oceanic suites and the progressively greater differences of the degree of silica saturation of late lavas, especially those of felsic composition, have been explained in terms of splitting and vertical spreading of the zone of melting and differentiation (McBirney, 1967). Such a mechanism seems required to explain compositional variations of the Tahitian rocks, which cannot be related to associated basic magmas in terms of the minerals found in the intrusive rocks or as phenocrysts in lavas (McBirney and Aoki, 1968). The composition of material that must be fractionated to produce the two Tahitian series is inconsistent with any plausible mineral assemblage likely to be stable in such liquids at shallow depths but fits remarkably well the assemblages that would be in equilibrium with an alkali basalt at depths corresponding to 10 and 20 kb.

Such evidence, even though supported by geological relations, experimental results, and current concepts of crystal fractionation, is still only permissive. The fact that alkaline gabbros and teschenite sills are known to differentiate to syenitic compositions at very shallow depths shows that deep-seated differentiation cannot be the sole mechanism by which strongly silica-deficient felsic rocks are produced.

One of the most intriguing features of the oceanic igneous suites is the manner in which the end products of each series reflect in magnified form the character of the basalt from which they are presumably derived. Silica-poor basalts produce differentiates that are decidedly more strongly undersaturated in silica, and silica-rich basalts produce felsic differentiates that are greatly enriched in silica. There is no discernable tendency, however, for the differentiates to fall into two extremities, as would be expected if they were trending toward distinct eutectics in a residua system. Instead, there is a complete spectrum between rhyolitic and phonolitic compositions, and if there is any composition of differentiates that is more common between these extremes than others, it is trachyte with a few percent of normative quartz.

When one contrasts the great diversity of mineral assemblages and conditions of crystallization with the broad similarities of differentiation trends for parental magmas of similar composition, it is difficult to believe that the nature of the crystallizing phases plays a role as important as that imposed by the inherent nature of the parent magma. Minerals crystallizing under equilibrium conditions reflect the nature and condition of the liquid in which they form but may not necessarily control or even substantially alter it. In many cases, their role must be a passive one, quite subordinate to other mechanisms as yet impossible to evaluate because of our inadequate understanding of the liquids themselves.

References Cited

Aoki, K., 1966, Phenocrystic spineliferous titanomagnetites from trachy-andesites, Iki Island, Japan: Am. Mineralogist, v. 51, p. 1799-1805.

Bailey, E. B., Clough, C. T., Wright, W. B., Richey, J. E., and Wilson, J. V., 1924, Tertiary and post-Tertiary geology of Mull: Scotland Geol. Surv. mem.

Bandy, M. C., 1937. Geology and petrology of Easter Island: Geol. Soc. America Bull., v. 48, p. 1589-1610.

Banfield, A. F., Behre, C. H. Jr., and St. Clair, David, 1956, Geology of Isabela (Albemarle) Island: Geol. Soc. America Bull., v. 67, p. 215-234.

Bates, H. W., 1892, Naturalist on the River Amazon: New York, D. Appleton & Co.

Best, M. G., and Mercy, E. L. P., 1967, Composition and crystallization of mafic minerals in the Guadalupe igneous complex, California: Am. Mineralogist, v. 52, p. 436-474.

Bott, M. H. P., 1965, Formation of oceanic ridges: Nature, v. 207, p. 840-843.

Brown, G. M., 1957, Pyroxenes from the early and middle stages of fractionation of the Skaergaard intrusion, East Greenland: Mineralog. Mag,. v. 31, p. 511-543.

Brown, G. M., and Vincent, E. A., 1963, Pyroxenes from the late stages of fractionation of the Skaergaard Intrusion, East Greenland: Jour. Petrology, v. 4, p. 175-197.

Bryan, W. B., 1967, Geology and petrology of Clarion Island: Geol. Soc. America Bull., v. 78, p. 1461-1476.

Bunsen, R., 1851, Uber die Processe der vulkanischen gesteinsbildungen Inseln: Poggendorff's Ann. d. Physik u. Chemie, v. 83, p. 223-224.

Carmichael, I. S. E., 1960, The pyroxenes and olivines from some Tertiary acid glasses: Jour. Petrology, v. 1, p. 309-336.

—— 1963, The occurrence of magnesian pyroxenes in porphyritic acid glasses: Mineralog. Mag., v. 33, p. 394-403.

—— 1964, The petrology of Thingmuli, a Tertiary volcano in eastern Iceland: Jour. Petrology, v. 5, p. 435-460.

Cavagnaro, David, 1965, Exploring the Galápagos on foot: Pacific Discovery, v. 18, p. 14-22.

Chesterman, C. W., 1963, Contributions to the petrography of the Galápagos, Cocos, Malpelo, Cedros, San Benito, Tres Marias, and White Friars islands: California Acad. Sci. Proc., v. 32, p. 339-362.

Chubb, L. J. 1930, Geology of the Marquesas Islands: B. P. Bishop Museum Bull. 68, 71 p.

—— 1933, Geology of the Galápagos, Cocos, and Easter islands: B. P. Bishop Museum Bull. 180, 67p.

Cox, Allan, and Dalrymple, Brent, 1966, Paleomagnetism and potassium-argon ages of some volcanic rocks of the Galápagos Islands: Nature, v. 209, p. 776-777.

Dall, W. H. and Ochsner, W. H., 1928, Tertiary and Pleistocene Mollusca from the Galápagos Islands: California Acad. Sci. Proc., v. 17, p. 89-139.

Darwin, Charles, 1891, Geological observations on volcanic islands: London, Smith, Elder & Co., 3d ed.

De Paepe, P., 1966, Geologie van Isla Daphne Mayor (Islas Galápagos): Natuurw. Tijdschr., v. 48, p. 67-80.

Durham, J. Wyatt, 1964, The Galápagos Islands Expedition of 1964: Am. Malacological Union, Ann. Rept, p. 53.

—— 1965, Geology of the Galápagos: Pacific Discovery, v. 18, p. 3-6.

Eibl-Eibesfeldt, Irenaus, 1959, Survey on the Galápagos Islands: UNESCO Mission Rept. no. 8, p. 12.

Engel, A. E. J., and Engel C. G., 1964, Igneous rocks of the East Pacific Rise: Science, v. 146, p. 477-485.

Ewing, M., Heezen, B. C., and Ericson, D. B., 1959, Significance of the Worzel deep sea ash: Natl. Acad. Sci. Proc., v. 45, p. 355-361.

Fisher, R. L., and Norris, R. M., 1960, Bathymetry and geology of Sala y Gomez, Southeast Pacific: Geol. Soc. America Bull., v. 71, p. 497-502.

Friedlander, I., 1918 Regelmässigkeit der Abstande vulkanischer Eruptionszentren: Zeitsch. f. Vulk., v. 4, p. 15-32.

Gast, P. W., Tilton, G. R., and Hedge, C., 1964, Isotopic composition of lead and strontium from Ascension and Gough islands: Science, v. 145, p. 1181-1185.

Granja, J. C., 1964, Geologia de la Isla Genovesa (Tower); Quito Ecuador, Editorial Universitario.

Green, D. H., and Ringwood, A. E., 1967, The genesis of basaltic magmas: Contr. Mineral. Petrol., v. 15, p. 103-190.

Green, W. Lowthian, 1887, Vestiges of the molten globe. Part 2: Honolulu.

Hess, H. H., 1949, Chemical composition and optical properties of common clinopyroxenes: Am. Mineralogist, v. 34, p. 621-666.

——1960, Stillwater igneous complex, Montana: Geol. Soc. America, Mem. 80, 230 p.

Kennedy, W. Q., 1931, On composite lava flows: Geol. Mag., v. 68, p. 166-181.

Kroeber, A. L., 1916, Floral relations among the Galápagos Islands: California Univ. Pubs. Botany, v. 6, p. 199-220.

Kuno, Hisashi, 1955, Ion substitution in the diopside-ferropigeonite series of clinopyroxenes: Am. Mineralogist, v. 40, p. 70-93.

Kuno, Hisashi, 1959, Origin of the Cenozoic petrographic provinces of Japan and surrounding areas: Bull. Volcanol., ser. II, v. 20, p. 37-76.

—— 1960, High-alumina basalt: Jour. Petrology, v. 1, p. 121-145.

—— 1962, Frequency distribution of rock types in oceanic, orogenic, and kratogenic volcanic associations: Am. Geophys. Union Pub. 1035, p. 135-139.

Kuschel, Guillermo, 1963, Composition and relationship of the terrestrial faunas of Easter, Juan Fernandez, Desventuradas, and Galápagos islands. California Acad. Sci. Occasional Paper no. 44, p. 79-95.

Kushiro, Ikuo, 1965, The liquidus relations in the systems forsterite-$CaAl_2SiO_6$-silica, and forsterite-nepheline-silica at high pressures: Carnegie Inst. Yearbook 64, p. 103-109.

Lacroix, Alfred, 1928, La constitution lithologique des iles volcaniques de la Polynesia Australe: Paris Acad. Sci. Mem., ser. 2, v. 59, no. 2.

—— 1936, Composition chimique des laves de l'Ile de Paques: Acad. Sci. Comptes Rendus, v. 202, p. 601-605.

Langseth, M. G., Grim, P. J., and Ewing, M., 1965, Heat-flow measurements in the East Pacific Ocean: Jour. Geophys. Research, v. 70, p. 367-380.

Laruelle, Jacques, De Paepe, P., and Stoops, G., 1964, Geologie de l'Ile Bartolome: Noticias de Galápagos, no. 4, p. 8-11.

Macdonald, G. A., 1960, Dissimilarity of continental and oceanic rock types: Jour. Petrology, v. 1, p. 172-177.

Macdonald, G. A., and Katsura, T., 1962, Relationship of petrographic suites in Hawaii, p. 187-195 *in* Crust of the Pacific Basin: Geophys. Mon. no. 6.

Macdonald, G. A., 1964, Chemical composition of Hawaiian lavas: Jour. Petrology, v. 5, p. 82-133.

Marshall, P., 1910, Note on the geology of Mangaia: New Zealand Inst. Trans., v. 42, p. 333.

McBirney, A. R., 1967, Genetic relations of volcanic rocks of the Pacific Ocean: Geol. Rundschau, v. 57, p. 21-33.

McBirney, A. R., and Aoki, K., 1968, Petrology of the island of Tahiti, p. 523-556 *in* Coats, R. R., Hay, R. L., and Anderson, C. A., *Editors,* Studies in volcanology: Geol. Soc. America Mem. 116 (Williams Volume).

McBirney, A. R., and Gass, I. G., 1967, Relations of oceanic volcanic rocks to mid-oceanic rises and heat flow: Earth and Planetary Sci. Letters, v. 2, p. 265-276.

Menard, H. W., 1964, Marine geology of the Pacific: New York, McGraw-Hill Book Co.

—— 1966, Fracture zones and offsets of the East Pacific Rise: Jour. Geophys. Research, v. 71, p. 682-685.

Menard, H. W., and Chase, T. E., 1965, Tectonic effects of upper mantle motion: Internat. Union Geol. Sci., Upper Mantle Symposium in New Delhi, 1964.

Muir, I. D., and Tilley, C. E., 1964, Basalts from the northern part of the rift zone of the Mid-Atlantic Ridge: Jour: Petrology, v. 5, p. 359-408.

Muir, I. D., and Tilley, C. E., 1966, Basalts from the northern part of the Mid-Atlantic Ridge, II, The Atlantis Collection near 30° N.: Jour. Petrology, v. 7, p. 193-201.

Nockolds, S. R., and Allen, R., 1954, The geochemistry of some igneous rock series, Part II: Geochim. et Cosmochim. Acta, v. 5, p. 245-285.

Noe-Nygaard, A., 1967, Variation in titania and alumina content through a three kilometer thick basaltic lava pile in the Faroes: Medd. Dan. Geol. For., v. 17, p. 125-128.

Nygren, W. E., 1950, Bolivar geosyncline of northwestern South America: Am. Assoc. Petroleum Geologists Bull., v. 34, p. 1998-2006.

O'Hara, M. J., 1965, Primary magmas and the origin of basalts: Scottish Jour. Geology, v. 1, p. 19-39.

—— 1968, The bearing of phase equilibria studies in synthetic and natural systems on the origin and evolution of basic and ultrabasic rocks: Earth Sci. Rev., v. 4, p. 69-133.

Peck, D. L., Wright, T. L., and Moore, J. G., 1966, Crystallization of tholeiitic basalt in Alae lava lake, Hawaii: Bull. Volcanol., v. 29, p. 629-655.

Peterson, M. N. A., and Goldberg, E. D., 1962, Feldspar distributions in South Pacific pelagic sediments: Jour. Geophys. Research, v. 67, p. 3477-3492.

Presnall, D. C., 1966, The joint forsterite-diopside-iron oxide and its bearing on the crystallization of basaltic and ultramafic magmas: Am. Jour. Sci., v. 264, p. 753-809.

Quensel, P. D., 1912, Die Geologie der Juan Fernandezinslen: Uppsala Univ. Geol. Inst. Bull., v. 11, p. 252-290.

Richards, Adrian, 1957, Volcanism in eastern Pacific Ocean basin: 1954-1955. Proc. XX Internat. Geol. Congress, Mexico, v. 1, p. 19-31.

—— 1958, Transpacific distribution of floating pumice from Isla San Benedicto, Mexico: Deep-Sea Research, v. 5, p. 29-35.

—— 1962, Active volcanoes of the Archipelago de Colon (Galápagos). Part XIV, Catalogue of the active volcanoes of the world: Naples, Internat. Assoc. Volcanology.

—— 1966, Geology and petrography of Isla San Benedicto: California Acad. Sci. Proc., v. 33, p. 361-414.

Richardson, Constance, 1933, Petrology of the Galápagos Islands: B. P. Bishop Museum Bull. 180, p. 45-67.

Robinson, B. L., 1902, Flora of the Galápagos Islands: Am. Acad. Arts and Sci. Proc., v. 38, p. 77-269.

Robson, G. R., Barr, K. G., and Luna, L. C., 1968, Extension failure: an earthquake mechanism: Nature, v. 218, p. 28-32.

Saha, Prasenjit, 1959, Geochemical and X-ray investigation of natural and synthetic analcites: Am. Mineralogist, v. 44, p. 300-313.

Shand, S. J., 1937, Earth-lore: London, Thos. Murby & Co., 2d ed.

Shumway, George, 1954, Carnegie Ridge and Cocos Ridge in the east equatorial Pacific: Jour. Geology, v. 62, p. 573-586.

Shumway, George, and Chase, T. E., 1961, Bathymetry in the Galápagos region: California Acad. Sci. Occasional Paper no. 44, p. 11-19.

Simkin, Tom, and Howard, K. A., 1968, The collapse of Fernandina caldera, Galápagos Islands: Smithsonian Institution Center for Short-Lived Phenomena, 12 p.

Slevin, J. R., 1931, Log of the Schooner ACADEMY on a voyage of scientific research to the Galápagos Islands, 1905-1906: California Acad. Sci. Occasional Paper no. 17.

—— 1959, The Galápagos Islands, a history of their exploration: California Acad. Sci. Occasional Paper no. 25, 150 p.

Smith, W. C., and Chubb, L. J., 1927, The petrography of the Austral and Tubuai Islands (southern Pacific): Geol. Soc. London Quart. Jour., v. 83, p. 317-341.

Stewart, Alban, 1911, A botanical survey of the Galápagos Islands: California Acad. Sci. Proc., ser. 4, no. 1, p. 7-288.

Stueber, A.M., and Murthy, V. R., 1966, Potassium-rubidium ratios in ultra-mafic rocks and the differentiation history of the upper mantle: Science, v. 153, p. 740-741.

Sutherland, F. L., 1965, Dispersal of pumice, supposedly from the 1962 South Sandwich Islands eruption, on southern Australian shores: Nature, v. 207, no. 5004, p. 1332-1335.

Tilley, C. E., 1950, Some aspects of magmatic evolution: Geol. Soc. London Quart. Jour., v. 106, p. 37-61.

Tilley, C. E., Yoder, H. S., and Schairer, J. F., 1965, Melting relations of volcanic tholeiitic and alkali rock series: Carnegie Inst. Yearbook 64, p. 69-92.

Upton, B. G., and Wadsworth, W. J., 1965, Geology of Réunion Island, Indian Ocean: Nature, v. 207, p. 151-154.

Vinton, K. W., 1951, Origin of life on the Galápagos Islands: Am. Jour. Sci., v. 249, p. 356-376.

Von Herzen, R. P., 1959, Heat-flow values from the southeastern Pacific: Nature, v. 183, p. 882-883.

Von Herzen, R. P., and Uyeda, S., 1963, Heat-flow through the eastern Pacific Ocean floor: Jour. Geophys. Research, v. 68, p. 4219-4250.

Wager, L. R., 1960, Types of igneous cumulates: Jour. Petrology, v. 1, p. 73-85.

Wakita, H., Nagasawa, H., Uyeda, S., and Kuno, H., 1967, Uranium, thorium and potassium contents of possible mantle materials: Geochemical Jour., v. 1, p. 183-198.

Walker, F., and Davidson, C. F., 1936, A contribution to the geology of the Faeroes (Denmark): Royal Soc. Edinburgh, Trans. v. 58, p. 869.

Washington, H. S. and Keyes, M. G., 1927, Rocks of the Galápagos Islands: Washington Acad. Sci. Jour., v. 17, p. 538-543.

Whitaker, T. W., and Carter, G. F., 1954, Oceanic drift of gourds—experimental observations: Am. Jour. Botany, v. 41, p. 697-700.

Wilcox, R. E., 1959, Universal stage accessory for direct determination of the three principal indices of refraction: Am. Mineralogist, v. 44, p. 1064-1067.

Williams, Howel, 1933, Geology of Tahiti, Moorea, and Maiao: B. P. Bishop Museum, Bull. 105.

Willis, Bailey, and Washington, H. S., 1924, San Felix and San Amrosio: their geology and petrology: Geol. Soc. America Bull., v. 35, p. 365-384.

Wilson, J. Tuzo, 1963, Continental drift. Sci. American, v. 208, p. 86-100.

Winchell, H., 1947, The Honolulu series, Oahu, Hawaii: Geol. Soc. America Bull., v. 58, p. 1-48.

Wolf, Teodoro, 1895, Die Galápagos-Inseln: Erde, v. 22, p. 246-265.

Yoder, H. S., and Tilley, C. E., 1962, Origin of basalt magmas: Jour. Petrology, v. 3, p. 342-532.

Author Index

193

Subject Index

Abingdon (Pinta) Island, 29, 71, 84, 88-94, 97, 107, 122
Academy Bay, 13, 33, 35, 38, 80
Aeolian Cove, 17, 19
Age determinations, radiometric, 19, 106
Albemarle (Isabela) Island, 4, 5, 30, 33, 54-72, 77, 97, 99, 101, 102, 107, 110, 119-121, 124, 130, 131, 140
Alcedo Volcano, 55, 56, 58, 59, 73, 102, 113, 121, 147
Alae lava lake, Hawaii, 153
Amphibole, 161-162
  analysis, 162
Ascension Island, 132

Bainbridge Rocks, 51
Baltra Island, 9, 12, 14, 17-20, 46, 97, 101, 105, 107, 177, 118
Barcena Volcano, Mexico, 114
Barrington (Santa Fe) Island, 9, 11-14, 15, 33, 39, 46, 52, 97, 101, 117-119, 160
Bartholomew Island, 51
Basalt
  analyses, 118, 121, 122-125
  feldspar-phyric, 92-94, 136-137
  principal types, 119-130
  submarine, 9-12, 13-17, 117-118, 169-170
Beagle Cone, 56, 60, 62-64, 72, 130, 133
Bindloe (Marchena) Island, 21, 29, 77, 84-88, 97, 107, 113, 122
Biological dispersal, 113-116
Biotite, 161
Boobies, 143
Brito-Icelandic province, 166
Buccaneer (Fresh Water) Bay, 4, 47, 53, 123, 130, 131, 144, 146, 150

Cabo Marshall, 68
Calderas
  Duncan Island, 37-38
  Albemarle Island, 56-61, 64-67
  Narborough Island, 73-83
  Bindloe Island, 85-87
  Wenman Island, 94-95
Caldwell Island, 26
Cape Berkeley, 56- 63, 68-71, 99, 102, 121
  Cowan, 46
  Ibbetsen, 90
  Trenton, 48
Carnegie Ridge, 55-57
Caroline Islands, 114
Cerro
  Azul, 55-57
  Brujo, 29
  Chivo, 29
  Colorado, 13-16, 106
  de los Gemelos, 25-26
  de Pajas, 23-25
  Inn, 48, 50, 52
  Mundo, 29
  Patricia, 29
Champion Island, 26
Charles (Florena) Island, 21-28, 33, 76, 77, 97, 99, 106, 123, 126, 130-133, 158
Chatham (San Cristobal) Island 3, 4, 5, 12, 21, 28-31, 99, 101, 107, 124, 127, 140, 166
Clipperton fracture zone, 169
Cocos Island, 166, 167
Cocos Ridge, 7, 98, 105, 109
Conway Bay, 16, 33, 103
Cormorant Point, 21, 25
Costa Rica, 7, 105, 109, 114, 168

195